Photoelectrochemical Water Splitting
Materials and Applications

Edited by

Inamuddin[1,2,3], Rajender Boddula[4], Mohammad Faraz Ahmer[5] and Abdullah M. Asiri[1,2]

[1]Chemistry Department, Faculty of Science, King Abdulaziz University, Jeddah 21589, Saudi Arabia

[2]Centre of Excellence for Advanced Materials Research, King Abdulaziz University, Jeddah 21589, Saudi Arabia

[3]Department of Applied Chemistry, Faculty of Engineering and Technology, Aligarh Muslim University, Aligarh-202 002, India

[4]CAS Key Laboratory of Nanosystem and Hierarchical Fabrication, National Center for Nanoscience and Technology, Beijing 100190, PR China

[5]Department of Electrical Engineering, Mewat College of Engineering and Technology, Mewat-122103, India

Published by **Materials Research Forum LLC**
Millersville, PA 17551, USA

Published as part of the book series
Materials Research Foundations
Volume 71 (2020)
ISSN 2471-8890 (Print)
ISSN 2471-8904 (Online)

Print ISBN 978-1-64490-072-7
eBook ISBN 978-1-64490-073-4

Distributed worldwide by

Materials Research Forum LLC
105 Springdale Lane
Millersville, PA 17551
USA
https://www.mrforum.com

Manufactured in the United States of America
10 9 8 7 6 5 4 3 2 1

Table of Contents

Preface

Production of fuels and chemicals from daylight offers a promising technology for sun-driven energy. Natural photosynthesis is the best-known paradigm for transforming solar energy into chemical energy. Our present energy sources are mostly from petroleum derivatives (oil, coal, and gas). The oil supplies seem to be drained in 50-150 years. The utilization of the non-renewable energy sources has brought about various ecological issues; for example, the expanding outflows of ozone layer harming substance, most outstandingly carbon dioxide (CO_2). The outflow of the ozone layer harming subtances seems to increase about two fold by 2050. Considering significant factors including the economy, the earth, and human wellbeing, the expanded measure of energy must originate from inexhaustible and sustainable sources. Among the majority of the sustainable sources (solar, wind, geothermal, tidal, and biomass energies), sun-powered energy is the most practical and appealing electrical energy source. Photoelectrochemical water splitting using solar energy is the technique used for the direct transformation of sunlight into hydrogen energy (H_2). The practical hydrogen energy production through photoelectrochemical water splitting requires perfect photoelectrochemical devices with solar-to energy conversion efficiency, low-cost and long-term stability. As of late, many endeavors have been made to achieve these objectives by photochemists, engineers, and scientific experts, working in the area of energy and environmental science. However, the acquired outcomes are still extremely constrained and not yet sufficient for practical applications. Hence, investigating the advanced photoelectrocatalyst materials are the basic need for fabricating efficient photoelectrodes for photoelectrochemical water splitting devices.

This book aims to deliver cutting-edge and up-to-date research findings on photoelectrochemical water splitting and how to achieve electrode materials characteristics and significant enhancements in electrochemical, physical, and chemical, properties of photochemical electrode and electrode materials. The central aim of this book is to present the recent developments in the area of photoelectrochemical water splitting. The chapters are written by leading experts, including researchers, academicians, and industrialists working in the field of energy conversion and storage. It provides indepth knowledge on fundamental aspects, key factors, mechanisms, theoretical insights, and experimental evidence for photoelectrochemical water splitting. The chapters discuss the various types of photoelectrocatalysts, i.e., noble metals, earth-abundant metals, carbon, polymer semiconductors and their synthesis methods, controlling bandgap, engineering strategies, and characterization techniques are also discussed in details. This book is bringing together contributions from leading researchers in academia and industry through out the world. It is a well structured and essential resource for scientists, undergraduate, postgraduate students, faculty, R&D professionals, energy chemists, and industrial experts.

Inamuddin[1,2,3], Rajender Boddula[4], Mohammad Faraz Ahmer[5] and Abdullah M. Asiri[1,2]

[1]Chemistry Department, Faculty of Science, King Abdulaziz University, Jeddah 21589, Saudi Arabia

[2]Centre of Excellence for Advanced Materials Research, King Abdulaziz University, Jeddah 21589, Saudi Arabia

[3]Department of Applied Chemistry, Faculty of Engineering and Technology, Aligarh Muslim University, Aligarh-202 002, India

[4]CAS Key Laboratory of Nanosystem and Hierarchical Fabrication, National Center for Nanoscience and Technology, Beijing 100190, PR China

[5]Department of Electrical Engineering, Mewat College of Engineering and Technology, Mewat-122103, India

Materials Research Forum LLC
https://doi.org/10.21741/9781644900734-1

Chapter 1

Transition Metal Chalcogenides for Photoelectrochemical Water Splitting

M.B. Costa, M. Medina, M.A.S. Andrade Jr.*, D. Coelho, L.H. Mascaro

Department of Chemistry, Federal University of São Carlos, Rod. Washington Luis, Km 235, ZIP 13565-905, São Carlos, Brazil

*marcos_asaj@hotmail.com

Abstract

Metal transition chalcogenide materials for photoelectrochemical water splitting have been assessed in this chapter. The structure and electrical properties of binary chalcogenides and the recent advances on chalcopyrite photocathodes are discussed, including the highest photoactive materials for this reaction, which are the copper-based chalcopyrite ($CuInS_2$ and $CuInGa(S,Se)_2$) and the kesterite ($Cu_2ZnSn(S,Se)_2$). Moreover, a brief review on silver-based chalcogenides as promising materials for photoelectrochemical water splitting application is presented.

Keywords

Transition Metal, Binary Chalcogenides, Chalcopyrite, Kesterite, Water Splitting

Contents

Photoelectrochemical Water Splitting: Materials and Applications   Materials Research Forum LLC
Materials Research Foundations **70** (2020) 1-42   https://doi.org/10.21741/9781644900734-1

1.   Introduction

According to the IUPAC recommendations on the nomenclature of inorganic chemistry with respect to collective names, the group 16/VIA of the periodic table is called chalcogens, i.e., the elements oxygen, sulfur, selenium, tellurium and polonium. Although Werner Fischer proposed the term in 1930 and Heinrich Remy recommended its official use in 1938 to the Reform of the Nomenclature of Inorganic Chemistry, the origin of this term remains obscure. The term chalcogen by William B. Jensen can be understood as "ore formers". The original Greek term *chalkos* means "copper" or "brass" and is still used in mineralogy and geochemistry to represent minerals containing copper in its structure [1,2].

Chalcogens have predominance of non-metallic character, exhibiting a tendency in increasing the metallic character in the heavy elements. As a result, O, S and Se are considered non-metals, while Te is considered a metalloid and Po a metal. The elements of this group have electronic configuration ns^2np^4, so that two more electrons are needed to keep its valence shell filled. As a consequence, the compounds formed from these elements are called chalcogenides [3]. Oxygen, like the rest of the elements of the second period of p-block, has significantly different properties from those of the rest of the group. The reason for that is the first element maximum coordination number of each group is four, unlike the other elements can vary between five, six or seven. The increase in the size of the atom is intrinsically linked to this phenomenon, which is possible to use its d orbitals in the bonds [4]. As a contextual consequence, some textbooks adopt this concept and treat it as a distinct chemical study: oxides.

Briefly, some of the major differences between the atomic properties of the oxygen atom and the chalcogens (S, Se, and Te) are listed below [5]:

a) S, Se, and Te atoms are heavier and larger than the oxygen atom;

b) The oxygen atom does not have d orbitals accessible in its configuration while those of S, Se and Te have 3d, 4d and 5d orbitals, respectively, making it more polarizable as down in the group;

c) S, Se and Te atoms are more electropositive than the oxygen atom, implying a more covalent when bound to a transition metal;

In this chapter, the term transition metal chalcogenides will be employed to sulfide, selenide and telluride-containing compounds in their structure along the transition metal as a distinction from their respective oxides.

Currently, the use of semiconductor materials for the solar harvesting is applied in three important areas: photocatalytic systems, which have important applications in photocatalytic water splitting and organic synthesis, and in the photoreduction of CO_2 and N_2; photovoltaic solar cells, which directly convert sunlight into electricticity; and photoelectrochemical cells, for example dye-sensitized liquid-junction and nanoparticle-sensitized liquid-junction solar cells and finally water splitting solar cell [6], where the materials based on transition metal chalcogenides will be discussed in this chapter.

Since the work of Fujishima and Honda [7] brought light to the electrochemical photolysis of water using a TiO_2-based electrode in 1972, many efforts have been done in the area of semiconductor materials applied to more efficient photoelectrochemical cells. This led to a continue search for semiconductors that exhibit rapid charge transport in the solid/solution interface and stable for long periods, in addition to an efficient absorption in a broad range of the visible spectrum. Tributsch is a pioneer in the use of transition metal chalcogenides as electrodes for photoelectrochemical solar cell, using natural p-conducting MoS_2 layer crystals and later with doped $MoSe_2$ [8,9]. Although work with CdTe and ZnTe had already been reported earlier, the application of transition metal chalcogenides as a photoelectrode in PEC water splitting was only extensively investigated years later [10].

It is worth mentioning that this section summarizes several transition metals chalcogenides-based materials used as photoelectrodes for water splitting, as well as in the presence of catalysts used for hydrogen evolution reaction (HER) or oxygen evolution reaction (OER) or even in the formation of heterojunctions that contribute to the photoelectrocatalytic performance. Theoretical studies that demonstrate interesting electronic properties for water splitting, such as suitable band gap energy and the most appropriate band-edge level alignment, will be also mentioned in this chapter to introduce new materials among those widely investigated. The compounds will be divided into binaries and multinaries (which includes one the highest efficient materials for photoelectrochemical water splitting, the CIGS).

2. Typical structures of transition metal chalcogenides

Transition metal chalcogenides (TMCs) present various stoichiometries and structures for binary and ternary compounds. For binary compounds, stoichiometries may vary from mono, di, tri and tetrachalcogenide to those with metal-rich or chalcogenide-rich phases [11]. Their crystalline structures can range from the three-dimensional (whose most common structures are cubic type NaCl, zinc blende, hexagonal NiAs or wurtzite) to the two-dimensional layered structures, where the CdI_2 type structure is most common [12]. For the ternary compounds, the most common structures are chalcopyrite, orthorhombic and kesterite [13]. However, due to the great advances made by layered structures and the pyrite-type transition metals dichalcogenides in solar energy conversion for water splitting, it is worth to mention their structural arrangements for a better understanding of the materials that are highlighted in this chapter. Other structures will also be discussed as mentioned.

The ascension of the 2D structure of graphene, and its subsequent technological advances on its intrinsic properties absent in the bulk, have promoted the development of ultrafine semiconductors that have been widely investigated in current days. The dimensional arrangement of the atoms in the structure is fundamental to the properties of the material, within several areas of knowledge [14]. However, graphene is a semimetal whose band gap is zero, which restricts its applications in areas of optoelectronic devices and boosting new technologies based on TMCs with more appreciable band gap [15,16].

Layered TMCs semiconductors are high stable binary compounds formed from layered crystalline structures with transition metals ranging from 4/IVB to 7/VIIB and some of the compounds from 9/VIIIB to 10/VIIIB as shown in Figure 1 [17]. The layers of the transition metal dichalcogenides (TMDCs) with general formula MX_2 – wherein M is a transition metal and X is a chalcogen (S, Se or Te) – interact by weak van der Waals forces between them, which turn out to allow its exfoliation in a few or even monolayers. The planes are distributed in the structure by means of the X–M–X configuration, with the chalcogen atoms in two hexagonally packed planes over a plane of cations [18].

Typically, each of these layers has thicknesses of ca. 6–7 Å. The strong covalent character of the M–X bond and the polarizability of the anions make these structures different from others with anionic elements, for example, the oxides. In order to fill the binding states of the TMDCs from the X–M–X arrangement, each metal atom must provide two electrons for each chalcogen, resulting in +4 and -2 oxidation states, respectively. Moreover, greater stability is conferred on the layers in relation to environmental species because this structure does not have dangling bonds on its surface, considering each chalcogen atom with a single electronic pair [14,19].

The metallic atom may have octahedral or trigonal prismatic coordination depending on the degree of ionicity of the M–X bonds. For octahedral coordination, a more ionic character of the bond leads to larger distances between the chalcogen ions, reducing the electrostatic repulsion between them; otherwise, prismatic trigonal coordination exhibits greater covalent bond character caused by the strong overlapping between the metal and the chalcogen atom orbitals, decreasing the distances between them [19].

Photoelectrochemical Water Splitting: Materials and Applications   Materials Research Forum LLC
Materials Research Foundations **70** (2020) 1-42   https://doi.org/10.21741/9781644900734-1

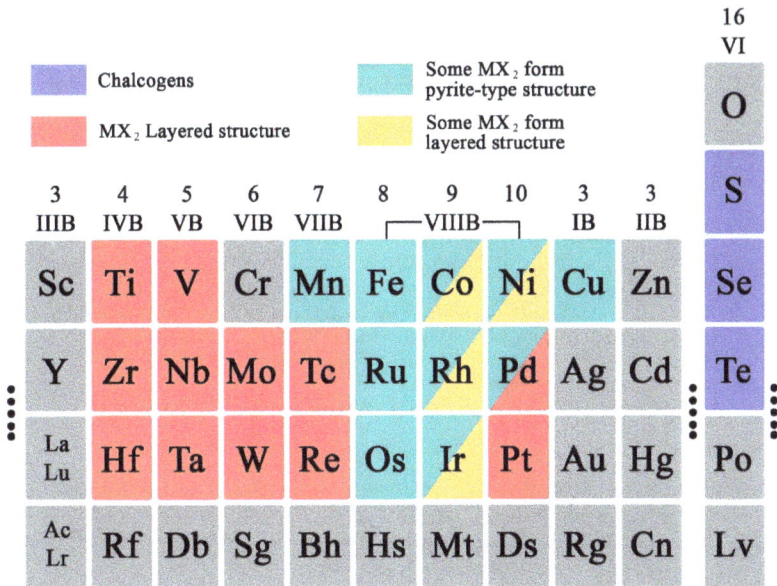

Figure 1: TMDCs compounds with layered and pyrite-like structures.

Several polytypes can be formed from the displacement between adjacent layers. Moreover, the polytypes of TMDCs bulk structures are supposed to depend on the transition metal which may exist in the form of tetragonal symmetry (1T, AA stacking and octahedral coordination of metal), hexagonal (2H, ABAB stacking and trigonal prismatic coordination of metal) and rhombohedral (3R, ABCABC stacking and prismatic trigonal coordination of metal) – the digit next to the letters T, H and R indicates the number of units X–M–X in the unit cell [20,21]. All that information is briefly described in Figure 2.

Generally, TMDCs present band gap energy between 0 and 2 eV, whose variation may be due to a series of factors, such as, the elements that compose the structure, the number of layers and the presence of doping atoms. This wide range of band gap energy is important for the applications of TMDCs, for example in the areas of catalysis, electrochemistry, thermoelectric, electronics, photochemistry and optical applications [22].

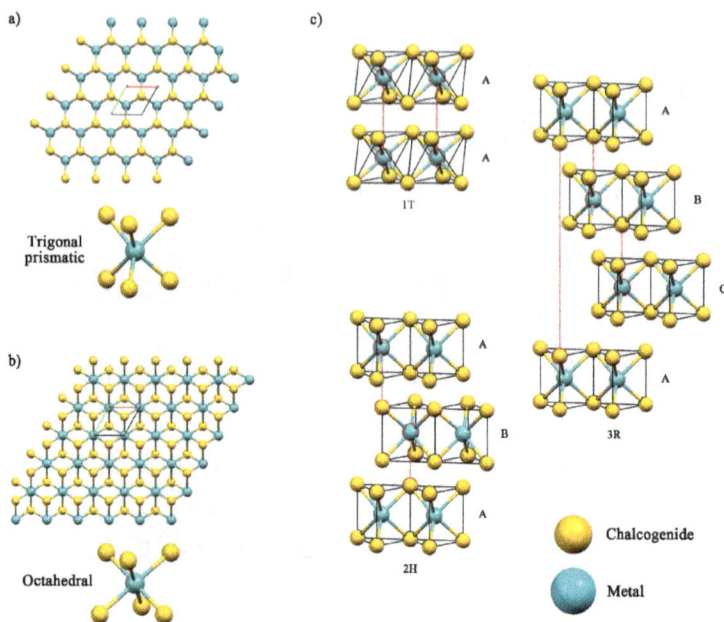

Figure 2: a) Schematic representation of the c-axis for a typical MX2 structure with trigonal prismatic and b) octahedral metal coordination. c) Scheme for different TMDCs polytypes: 1T, 2H and 3R with layers AA, ABA and ABCA stacked, respectively. The chalcogenides atoms are represented by yellow spheres and the transition metals by blue spheres.

Another important type of TMDCs is the pyrite-type three-dimensional structure (Figure 3). Its crystalline system is cubic and very similar to the NaCl rocksalt structure, where the Na^+ atoms are replaced by those of the transition metal, and Cl^- is replaced by the ion pair X_2^{2-}. The pyrite-type structure is possible to form due to the anionic pair X_2^{2-}. The chalcogen atoms pairing is a result of the increase of binding energy of the valence d-orbitals of the metal. Because of the formation of a three-dimensional arrangement, pyrite-like structures are isotropic in all directions. Although the marcasite-type structures can also be formed in the TMDCs, these structures are more unstable and less common in photoelectrochemical applications due to their low band gap energy values [19].

Photoelectrochemical Water Splitting: Materials and Applications   Materials Research Forum LLC
Materials Research Foundations **70** (2020) 1-42   https://doi.org/10.21741/9781644900734-1

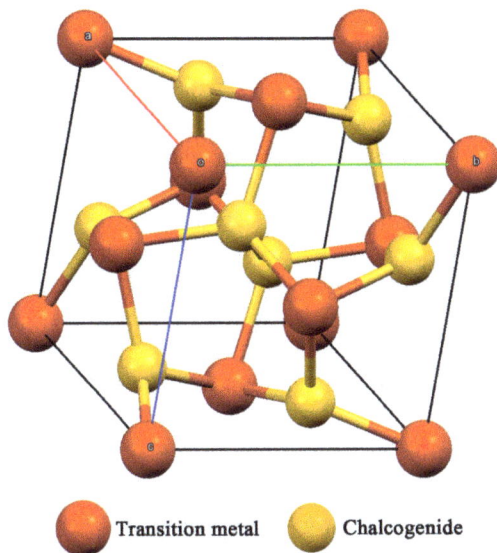

| Transition metal | Chalcogenide |

Figure 3: Crystal structure for pyrite-type MX2. The chalcogenides atoms are represented by yellow spheres and the transition metals by red spheres.

3.   Binary chalcogenides applied to photoelectrochemical water splitting

Titanium-based sulfide compounds TiS_3 have shown great technological interest as a low-toxicity and inexpensive semiconductors whose applications are in diverse areas such as thermoelectric devices, hydrogen storage and battery cathodes [23]. More specifically, titanium trisulfide (TiS_3) has recently been considered for PEC water splitting as a photoelectrode for hydrogen production. TiS_3, as the other 4/IVB group trichalcogenides, has the monoclinic $ZrSe_3$-type structure formed by infinite chains of TiS_6 unit stacked parallel to b-axis. These are of particular interest because of their structures offering a quasi-1D crystalline anisotropy formed from thin nanoribbons, in addition to exhibiting high electronic mobility and low resistivity [24,25]. The material has shown two different direct band gaps between 1.1 eV and 1.6 eV, obtained from photocurrent spectral response measurements and tauc plot [23]. More recently data demonstrate that TiS_3 is an n-type semiconductor that has shown high photocurrent results under white light illumination, suitable position for water reduction and photogenerated hydrogen rate of 1.80 μmol H_2 min^{-1} at 0.3 V vs. Ag/AgCl, achieving 7% photoconversion efficiency [24].

Niobium trisulfide NbS_3 is a semiconductor that exhibits band gap conductivity ranging from 0.60 to 1.01 eV. Its structure is triclinic, formed by chains of trigonal prisms of one dimension with the metallic atom near the center. NbS_3 is a transition metal trichalcogenide with formula MX_3, which exhibits electronic properties like Peierls transition and formation of waves of density of charge [26,27]. Its use as a photoelectrode is very scarce in the literature, presenting better results when combined with titanium, forming a ternary compound of the formula $Nb_xTi_{1-x}S_3$ (titanium rich, monoclinic) or $Ti_xNb_{1-x}S_3$ (niobium rich, triclinic). Among them, $Nb_xTi_{1-x}S_3$ presented the best results of photogeneration hydrogen rate, with average flux of 2.2 μmol H_2 min^{-1} cm^{-2}, i.e, a value three times greater than the binary NbS_3 and 25% higher than the TiS_3. The photoanodes yield efficiencies of $Nb_xTi_{1-x}S_3$ corresponded to a value 3.5 times higher than NbS_3 and 80% greater than TiS_3 [26].

Zirconium-based chalcogenides compounds ZrS_2, $ZrSe_2$ and ZrSSe have not been much exploited for photoelectrochemical application. Zr is a non-toxic and the seventh most abundant element of the earth. Specifically, ZrS_2 has a CdI_2 type layered structure with 1T polytype stacking which interacts between them by weak van der Waals forces, as well as all 4/IVB dichalcogenides. These structures are very similar to those of MoS_2 and WS_2, differing only by octahedron arrangement of the zirconium cation compared to the prismatic trigonal of the MoS_2 and WS_2 cation. ZrS_2 has an indirect E_g of 1.68 eV, i.e., as expected according to theoretically obtained. A ZrS_2 single crystal behaves as an n-type semiconductor when used as an electrode in a PEC cell. Although ZrS_2 does not appear to be a material of great interest for conventional photoelectrochemical cells, its electrochemical properties can be exploited in intercalation compounds for energy storage [28]. Other Zr-chalcogenides single crystals, such as $ZrSe_2$ and ZrSSe, were also investigated in PEC applications, but no information is available in the literature regarding their photoactivity. These materials exhibit n-type conductivity and have a band gap of approximately 1.69 eV for $ZrSe_2$ and 1.46 eV for ZrSSe. However, the positioning of the energy bands of the materials revealed that these may not be the most suitable for water splitting [29].

Molybdenum disulphide MoS_2 is an inexpensive, earth-abundant, non-toxic and excellent semiconductor to replace catalysts based on precious metals in the development of photoelectrochemical cells. Although it has been widely used as an electrocatalyst for HER due to several improvements to the system, its photocatalytic activity is negligible and its direct use as a light-harvesting was rarely reported in the literature [30,31]. Its structure consists of S–Mo–S sheets interacted by van der Waals forces, and can assume four different polytypes, i.e., 1H MoS_2, 1T MoS_2, 2H MoS_2 and 3R MoS_2, of which the 1H phase is the most stable between the others and the 1T and 3R are meta-stable [32]. Bulk MoS_2 is presented as an indirect band gap semiconductor with a value of approximately 1.3 eV while MoS_2 monolayers has a E_g of ~1.9 eV [33,34].

PEC measurements employed in MoS_2 photocatalysts have shown interesting results when combined with reduced graphene oxide (rGO). MoS_2 nanoparticles were deposited on rGO, indicating that the MoS_2/rGO system provides more available catalytically edge sites than large aggregates of pristine MoS_2, resulting in a higher photocatalytic activity under visible light [35]. Thin nanoplatelet structures were also fabricated from *p*-type MoS_2 grown on *n*-type rGO doped with nitrogen to form numerous nanoscale *p-n* junction. The *p-n* junctions in the nanoplatelet structure resulted in a positively shift onset potential, an efficient separation of charge and an improvement in the generation of electron-hole pairs, increasing the PEC activity for the hydrogen production when compared to the bare MoS_2. In the *p*-MoS_2/*n*-rGO system, while MoS_2 acted as a catalytic and photocatalytic center for absorption of the solar spectrum, N-doped *n*-rGO acted as an active element in heterostructures [30]. When applied as photoelectrode in PEC water splitting, researchers obtained a photoanode composed of Au nanoparticles grown on MoS_2 nanosheets using FTO as substrate. The single layers of MoS_2 were prepared using controllable lithium exfoliation method while the gold nanoparticles were grown from a 0.3 mM $HAuCl_4$ aqueous solution. Subsequently, Au-MoS_2 composites were spin-coated on FTO using the suspension in methanol. It was possible to observe that the plasmon induced effect of Au nanoparticles grown on MoS_2 nanosheets greatly improved the photocatalytic activity for water splitting. The Au-MoS_2 composite presented a 370 μA cm^{-2} photocurrent at 0.8 V vs. Ag/AgCl, while pristine MoS_2 exhibited a value of 100 μA cm^{-2} at the same potential [31].

Recently, researchers fabricated a bilayer heterojunction from the monolayer *n*-type MoS_2 semiconductor coupling with *p*-type WSe_2, aiming to improve photocatalytic efficiency for PEC water splitting. The monolayers of MoS_2 and WSe_2 were stacked layer-by-layer during the manufacturing of an ultrathin vertical heterojunction by microfabrication method. As a result, MoS_2/WSe_2 heterojunction increased the light-collection range and the separation of the electron-hole pair, demonstrating photocurrent of 28 μA cm^{-2} in 1 V, i.e., a much higher value than that of pure WSe_2 and MoS_2 [36].

Tungsten diselenide WSe_2 is a semiconductor with layered crystalline structure with 2H polytype, as well as other dichalcogenides (S and Se) of molybdenum and tungsten. Its bond is predominantly covalent, whose tungsten atom is in trigonal coordination with six other selenium atoms. Its electrical conductivity is higher due to the permitted anisotropy of its layer structure, if compared to the materials with electron transport perpendicular to the planes in which the electron travels through the hopping mechanism [12].

WSe_2 is a p-type semiconductor that shows great interest in photoelectrochemical and photovoltaic devices because it demonstrates excellent performance in the conversion of solar energy. WSe_2 showed highly efficient photoelectrochemical conversion, reaching a value of 14% by means of the passivation of the recombination centers using photoetching of the van der Waals faces of the WSe_2 [37]. WSe_2 has adjustable band gap according to number of atomic layers and high absorption of incident light (up to 10% absorption) for thicknesses < 1 nm. In fact, studies have shown that the band gap energy value decreases as it goes from sulfide to telluride and from molybdenum to tungsten. Otherwise, as the thickness of the bulk layer is reduced to monolayer, the blue-shifted indirect band gap occurs for the direct band gap of the monolayer [38,39]. As a consequence, bulk WSe_2 presents indirect band gap of approximately 1.20 eV while its monolayer displays a direct band gap value of 1.65 eV [40]. In addition, WSe_2 is anti-photocorrosive due to the d–d transition in its crystalline lattice, which turns out to be an attractive for PEC devices [41].

When applied in PEC water splitting, the material exhibits efficient solar-to-hydrogen conversion, reaching photocurrent up to 4 mA cm^{-2} and 60% internal quantum efficiency for a solution-processed TMDC semiconductor [42]. The result is obtained using a pre-annealed WSe_2 powder (with Se at 1100 °C for 8 hours) followed by powder exfoliation and subsequent surfactant treatment, in the presence of optimized Pt-Cu cocatalyst in samples. In summary, WSe_2 is very promising material for hydrogen production and a potential semiconductor for solar energy conversion.

Chromium monochalcogenides CrX (X = S, Se and Te) have recently been considered for potential photocatalytic applications. These belong to the new family of 2D pentagonal monolayers, whose material is semiconducting and antiferromagnetic. Its monolayer is formed by three atomic layers of Cr-X-Cr with tetragonal symmetry, unlike the pentagonal 2D monolayers are constituted of pentagonal building blocks. Based on computational methods, the material was found to be very stable against thermal and mechanical perturbations, as well as displaying band gap calculated at 2.38, 2.52 and 1.30 eV for CrS, CrSe and CrTe, respectively, which is a good indication for absorption of the visible spectrum employed in photocatalysis for water splitting. From the first-principles calculations, penta-CrX has the alignment of bands that allow proton reduction and water oxidation, except 2D penta-CrTe, which does not produces O_2 due to the energy positioning of its band valence to be less positive than that needed to oxidize water [43].

Manganese sulphide MnS has attracted interest from researchers due to its properties being assigned to various areas of interest, such as applications in solar cells as window/buffer material, energy storage and photocatalysis, as well as optoelectronic and luminescent devices, and Li-ion batteries. MnS is a *p*-type wide band gap semiconductor (3.0 – 3.7 eV) and its structure is characterized by three crystallographic phases. The α-MnS phase exhibits cubic structure of the rocksalt type and stability above ambient temperature while the β- and γ- phases exhibit crystalline structures of the zincblende and wurtzite, respectively, in addition to being metastable phases [44,45]. However, its wide band gap appears as a disadvantage to the material, which limits the light absorption and consequent efficiency of the device. Therefore, heterostructures are built to overcome this problem and improve their photocatalytic and photoelectrochemical performance [46].

Recent research findings have revealed that heterostructures consisting of MnS nanoparticles coupled to 2D ultrathin MoS_2 monolayers exhibited excellent improvements in photocatalytic and photoelectrochemical activities compared to pristine MnS. The positioning of the energy bands of both materials also resulted in a better separation of the charge carriers, allowing a more efficiently charge transport between the semiconductors [47]. PEC measures have been used to characterize the transient photocurrent response of core-shell structures in which MnS decorated surfaces of CdS nanorods synthesized by hydrothermal process. A slurry composed of CdS@MnS composites, carbon black, a binder and solvent N-methyl pyrrolidone was coated onto ITO plate working electrode were used. The sample was irradiated under visible light (300 W) and chopped at 0 V vs. Ag/AgCl. The transient revealed an increase photocurrent of ca. 17% for the composite compared to bare CdS. Additionally, photocatalytic H_2 evolution presented a rate of 15.55 mmol h^{-1} g^{-1} and an apparent quantum efficiency of 6.9%, demonstrating that the structure CdS@MnS benefits the photocatalytic hydrogen evolution [48].

Rhenium disulfide (ReS_2) belongs to the group of transition metal dichalcogenides with distorted 1T structure. The ReS_2 single layer primitive cell contains four Re atoms and eight sulfur atoms, whose two hexagonal planes of the sulfur atoms are intercalated by a hexagonal plane of the Re atoms [49]. The bulk ReS_2 has a triclinic structure with a hexagonal lattice via Peierls distortion [50]. Unlike other MX_2 chalcogenides, electronic and vibrational ReS_2 properties are minimally independent of the number of layers because of the poor interactions among these. ReS_2 is a semiconductor with a direct band gap of 1.35 eV for bulk and 1.43 eV for the monolayer, which can enable an efficient absorption of the visible spectrum and consequent performance in the conversion of solar energy [49,51]. In fact, this difference in band gap energy is small compared to the band gap of other TMDCs, where the band gap depends significantly on the number of layers

[51]. In addition, density functional theory calculations revealed that both monolayers and bulk ReS_2 did not show significant band gap changes and energetic positioning of the bands, which are appropriately suitable for photocatalytic water splitting [49,52].

ReS_2 nanosheets arrays have also been grown vertically on p-Si planar photocatalysts to act both as an efficient electrocatalyst for HER and semiconductor to form a type-II heterojunction. The *p*-Si/ReS_2 photocatalyst showing excellent stability and catalytic performance for PEC water splitting, exhibiting a photocurrent of 9 mA cm^{-2} (at 0 V vs. RHE) and conversion efficiency of 0.28% (half-cell STH), that were 23 and 35 times higher than *p*-Si, respectively. Both abilities of acting as a catalyst and semiconductor allow the bifunctional ReS_2 to be able to improve light harvesting, charge separation, and catalytic kinetics [53].

Nanosheets of layered ReS_2 grown on conductive carbon fiber clothes (CFC), have demonstrated *p*-type conductivity, favorable electron-band structure and highly active edge sites. ReS_2@CFC has presented natural coupling of photoelectric response and catalytic kinetics, thereby making the material a promise in PEC water splitting for use either as co-catalyst or photo activated catalyst [54].

Technetium-based dichalcogenides TcX_2 (X = S and Se) have been less investigated transition metal dichalcogenides. Recent studies from DFT show that these semiconductors in layers of triclinic structure exhibited indirect band gap of 1.91 and 1.69 eV for 2D TcS_2 and $TcSe_2$, respectively. Theoretical calculations have shown that these materials have much better light absorption capacity than MoS_2 and $ReSe_2$ monolayers, making them potential light absorbers. In particular, $TcSe_2$ displays the position of the energy bands suitable for photocatalytic water splitting [55].

Although iron disulfide FeS_2 is not as widely used as other TMDCs (e.g., MoS_2, WS_2, NiS_2, $NiSe_2$, CoS_2 and $CoSe_2$) for water splitting. However, there are few papers available in the literature reporting their photocatalytic activity. More specifically, the literature becomes scarce when using FeS_2 in photoelectrochemical cell for water splitting [12,56].

FeS_2 has a pyrite-like cubic structure and attractive optoelectronic properties for various purposes. such as high absorption coefficient $> 10^5$ cm^{-1} and narrow band gap, equivalent to 0.95 eV for indirect and 1.03 eV for direct band gap. The high carrier mobility and the long diffusion length of the minor carriers, 360 cm^2 V^{-1} S^{-1} and 0.1–1 μm, respectively, are also attractive for the material. Being an earth-abundant, non-toxic and inexpensive semiconductor, FeS_2 is an alternative among TMDCs for demonstrating high quantum efficiency and good stability against photocorrosion in PEC cells [57,58]. However, the material still faces some challenges in the conversion of solar energy that makes it far from its applicability, such as low OCP (< 0.2 V) [59] and narrow band gap.

FeS_2 nanoparticles were obtained by solvothermal method with high uniformity and good crystallinity, besides producing high photocurrent in the direction of hydrogen production under illumination by a 300 W Xe lamp (with a 390 nm filter) in neutral electrolyte. The slurry composed by the FeS_2 catalyst and 10 µL of 2.5% polyvinylidene fluoride (PVDF) N-methylpyrrolidone solution was coated onto ITO conductive glass. It was possible to note that smaller catalyst particles resulted in the higher photocurrent of 3.3 mA cm^{-2} at -0.80 V vs. NHE [60]. FeS_2 was also coupled with hierarchical structure top-porous-bottom-tubular TiO_2 nanotubes, demonstrating high photoelectrochemical performance in both the visible and ultraviolet and infrared regions. FeS_2/TiO_2 NT's showed photocurrent with three orders of magnitude greater than pristine TiO_2 NT's in the infrared region. This improvement is probably due to the narrow band gap of FeS_2 and to the electronic and morphological properties of TiO_2 [59].

There are many reports in literature on transition metals disulfides like MS_2 (M = Fe, Co, Ni and Mo) when used as electrocatalysts, however, few papers report their direct application to PEC water splitting. In fact, the stoichiometry, dimensionality and quantum size effect play a very important role in the adjustment of its electronic structure, which can significantly favor its photoelectrocatalytic performance [61]. Among these transition metal disulfides, NiS_2 is a Mott insulator with band gap between 0.3 and 0.8 eV and intrinsically metallic surface, i.e., it does not follow the conventional band theories due to the incompatibility of its band structure [62]. For this reason, nickel sulfide has recently been obtained with $NiS_{1.97}$ stoichiometry in photoelectrochemical applications for hydrogen catalysis via water splitting. The material was obtained by the sulphurization method of nickel oxide, which in turn was synthesized from the calcination of $Ni(OH)_2$ at 600 °C for 2 hours. Sulfur deficiency in the material led to changes in the density of electronic states. The band gap value was 1.32 eV, which can be attributed to the high concentration of sulfur vacancies. These defects also resulted in increased carrier density, which has a strong effect on pinning of Fermi level. Therefore, band alignment has become more favorable considering the water redox potential. As a result, a PEC performance of 1.25 mA cm^{-2} was obtained at 0.68 V vs. NHE in pH 7.0 (1000 times higher than stoichiometric NiS_2) [61].

Several noble-metal transition dichalcogenides based on Pt or Pd PdS_2, PtSSe and PtX_2 (X = S and Se) have been rarely investigated in the various fields of application due to their high costs. However, DFT theoretical calculations of electronic structure have shown that the compound PdS_2 is can be applied for water photocatalysis [63]. Studies have demonstrated that PdS_2 has typical layered structures with orthorhombic arrangement in bulk [64], with a most stable configuration for single layer of PdS_2, unlike traditional 2D TMDs. It is characterized with planar tetra-coordinate Pd atoms and covalent S–S bonds [65]. According to the investigations, from the single layer band structures were observed a band gap of 1.24-1.36 eV, but the functionalization of the single layer (replacing S by P, N and C) decrease its band gap. In addition, it can be observed that PdS_2 monolayers can exhibit catalytic properties for both HER (with N-functionalization) and OER (S-monovacancy defects) [63]. Therefore, this makes the use of the material interesting for possible applications in photoelectrochemical cells.

Regarding materials using Pt in its structure, the photocatalytic behavior of 2D Janus PtSSe has been investigated by first-principles calculations. The octahedral phase PtSSe monolayer exhibited Eg =2.19 eV, high extinction coefficient and appropriated band edge energy for water splitting. In addition, the high carrier mobility and the large internal electric field also contributed to photocatalytic application [66]. Using a first-principles design approach, it was possible to observe that monolayer of PtS_2 and $PtSe_2$ also exhibited band edge energy levels suitable for photocatalysis of water [67].

The chalcocite structure (Cu_2S) can occur in three different forms: the low chalcocite phase with monoclinic crystalline structure, the high chalcocite hexagonal structure and the chalcocite with cubic structure (Figure 4) [68]. Although stable Cu_xS compounds at temperatures lower than 90 °C ($Cu_{1.94}S$, $Cu_{1.8}S$, $Cu_{1.75}S$, in addition to Cu_2S itself) have a E_g = 1.1 – 1.2 eV, Cu_2S has E_g = 1.2 eV, all of them with p-type conductivity [68,69]. Moreover, the lack of Cu^+ in the djurleite phases ($Cu_{1.94}S$), digenite ($Cu_{1.8}S$) and anilite ($Cu_{1.75}S$) are compensated by large concentration of holes and consequent formation of a degenerate semiconductor, that is , unsuitable for conventional PV [70].

The initial studies on Cu_2S nanowires isolated for PEC water splitting applications were made from obtaining the material on copper foil using gas-solid reaction. Among the studied conditions, bare Cu_2S and Cu_2S/CdS nanowires were those that presented better results. These showed cathodic photocurrents with values around 0.11 e 0.14 mA cm^{-2} at KCl 0.1 mol L^{-1}, respectively, applying a bias of -0.50 V vs. Ag/AgCl [71]. In another case, the use of the gas-solid reaction was again used to obtain the Cu_2S nanowires array (NWAs) photocathodes onto Cu foil and the dip-assembly of carbon quantum dots (CQD) on Cu_2S. The system provided a photocurrent of 1.05 mA cm^{-2} at 0 V vs. NHE and photoconversion efficiency of 0.148%, i.e., a value four times higher than the non-modified Cu_2S NWAs. This improvement was attributed to the effects belonging to the CQD, related to the improvement of charge transfer and the energy-down-shift effect [69].

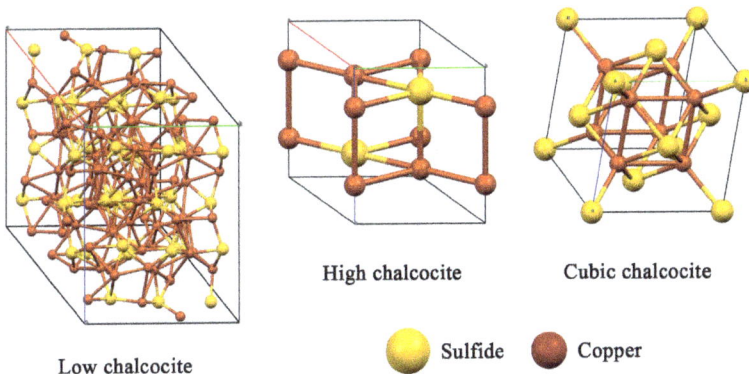

Figure 4: Crystal structure for different types of chalcocite (Cu_2S): the low chalcocite (monoclinic), the high chalcocite (hexagonal) and the chalcocite with cubic structure [49]. The sulfide atoms are represented by yellow spheres and the copper atoms by red spheres.

Photocathode of $Cu_{1.94}S$ nanoflake array (NFAs) was also reported for PEC water splitting with an effective strategy to confine abundant defects within the surface layer. It is known that surface defects in semiconductor materials can lead to high density of carriers and consequently cause localized surface plasmon resonance. Materials with different defect distribution, i.e., fully defective, superficial defective and defect-free $Cu_{1.94}S$ NFAs can be controlled from the synthesis. When observed in $Cu_{1.94}S$, the localized surface plasmon resonance phenomenon presented several advantages in the material with surface defects, such as favoring the light harvesting and the separation of charges, near-infrared absorption and surface reaction kinetics, exhibiting a photocurrent 0.597 mA cm^{-2} in 0 V vs. RHE. This demonstrates that control over defects in material plays an important role in achieving efficient solar energy conversion [72].

CdS/Cu_2S core-shell nanorods were obtained on FTO and employed as photoanode for PEC water splitting at pH maintained at 7.2 using 0.5 M Na_2SO_4 as supporting electrolyte and 0.2 M Na_2SO_3 as hole scavenger, in addition to Pt as counter electrode and Ag|AgCl/Cl$^-$ as reference electrode. The thinner shell of Cu_2S provided the highest ABPE (applied bias photon-to-current efficiency) conversion efficiency (6.79%) while the thicker shell provided a greater depletion region and consequently an increase in carrier recombination kinetics and lower efficiency (6.13%). The system presented excellent photoelectrocatalytic performance, providing photocurrent of 8.12 mA cm^{-2} for the thinner shell against 6.78 mA cm^{-2} using the electrode with the thicker shell and 3.05 mA cm^{-2} for the bare CdS in 0.5 V vs. RHE [73].

Cu_2S-coated Cu_2O nanowires (NWs) photocathodes were also obtained on porous copper foam (CF). Cu_2S thin films have several advantages for the system, such as increased of the absorption spectrum for the near infrared region, suitable band edge energy levels with Cu_2O, for a more effective electronic transfer and as a protection layer for Cu_2O. When illuminated, Cu_2S/Cu_2O NWs CF exhibited a photocurrent density of -5.05 mA cm^{-2} (at 0 V vs. RHE), which was 25% higher than Cu_2O NWs/CF, demonstrating superior PEC performance using Cu_2S-coated Cu_2O [74].

Recently, researchers have opted an indirect method for synthesis of the stoichiometric Cu_2S via solution-processed ion exchange reaction. Briefly, CdS is deposited on Au-coated FTO substrate via CBD and subsequently immersed in Cu^+ containing aqueous solution. Thus, the CdS films through ion exchange reaction, transformed completely into Cu_2S. The film was also coated with a protective layer of TiO_2 and RuO_x-based hydrogen catalysts. The photocathodes of $Cu_2S/CdS/TiO_2/RuO_x$ configuration exhibited an increased photocurrent of 7.0 mA cm^{-2} at a -0.3 V vs. RHE applied potential [75].

Besides these cited heterostructures, several other structures based on copper sulfide have already been applied to water splitting. Copper-based binary chalcogenides have been an excellent alternative for photocathodes demonstrating a low-cost material and a strong candidate for hydrogen production.

Silver sulfide (Ag_2S) has three polymorphic structures, namely: a stoichiometric low-temperature monoclinic phase (acanthite); and two non-stoichiometric phases, i.e. body centered cubic (argentite) and high-temperature face centered cubic phase (sometimes called high argenite) [12,76]. Ag_2S has a low band gap equivalent to 1.06 eV, whose value corresponds to the same of the bulk, which allows the material to absorb a larger range of the solar spectrum and to be attractive to couple it to the CdS [77,78].

Silver sulfide has already been used as a cocatalyst to improve the separation of charges in a heterojunction constituted by CdS/Ag_2S, in which Ag_2S was deposited via SILAR (successive ionic layer adsorption and reaction) on CdS nanorods. The insertion of Ag_2S as cocatalyst provided an improvement in the photoactivity of the system, presenting a photocurrent with values three times higher than the electrode formed from bare CdS [79].

Another system that has proven to be very efficient to the photoelectrochemical water splitting is the hierarchical graphene/CdS/Ag_2S sandwiched nanofilms. The graphene/CdS/Ag_2S nanofilms were obtained on ITO substrate following the combination of the dip-coating method (in the obtaining of graphene oxide), the SILAR technique (for coating with CdS and Ag_2S films) and subsequent reduction by cyclic voltammetry. This material has a E_g of 2.13 eV while the photocurrent density of the nanostructured system reached to a value above 5.18 mA cm^{-2} at 0.55 V vs. Ag/AgCl with maximum hydrogen generation of 221 μmol h^{-1} and efficiency of 4.11%. The high performance of graphene/CdS/Ag_2S sandwiched nanofilms can be attributed to several improvements, including solar spectrum absorption enhancement and consequent high light-harvesting promoted by the CdS/Ag_2S junction, to the proper alignment of the band-edge levels of the CdS/Ag_2S system and the improvement of the electron-injecting efficiency promoted by graphene. Again, the design of electrodes from hierarchical sandwiched nanofilms showed an efficient proposal for new strategy employed in PEC water splitting [78].

Cu_2ZnSnS_4-Ag_2S-based nanostructures as sensitizers have also been reported as absorber material for water splitting. Type II based nanoheterostructures have been obtained from the cation exchange with Ag^+ ions to form CZTS-Ag_2S nanoparticles of controlled composition. The formation of nanoscale *p-n* heterostructures have allowed a better energy band levels alignment for better separation of charges, leading to high photocurrent density (0.58 mA cm^{-2}) compared to the pure CZTS electrode (0.06 mA cm^{-2}) [80].

ZnS is an n-type semiconductor that can act as a photoelectrode in the hydrogen evolution reaction. It exhibits a rapid generation of photoexcited charge carriers and presents the conduction band in high negative potential that enables the rapid electron transfer. However, this material has a high band gap value of approximately 3.6 eV [100] which means that its activity is restricted to UV light absorption, corresponding only 4% of the total sunlight spectrum.

In this sense, ZnS photoelectrocatalysis studies have been involved in search of alternatives that can increase the sunlight absorption region of the photocathode. ZnS doping has been generally used for this purpose. In this context, it is known that the presence of Cu metal dopant improves the visible light absorption and In dopant can improve the photogenerated electrons migration. Based on it, the Cu and In co-dopant were applied to improve the ZnS photocatalytic activity. The adequate addition on ZnS was achieved by 4 mol% In and 4 mol% Cu resulting in 4In4CuZnS nanoparticles. The synthesized ZnS nanostructures have shown a cubic lattice phase and the In and Cu co-doped ZnS materials matched well with cubic In_2O_3 cubic structure. The photocatalytic activity was tested in 0.1 M NaS_2 solution at pH 3. The obtained photocatalyst was an n-type semiconductor as evident by the photoelectrochemical response at positive potentials, achieving a photocurrent density of 12.2 mA cm^{-2} at 1.1 V. The pristine ZnS and the 4In4CuZnS have shown maximum hydrogen evolution rates close to 100 and of

1189.4 μmol h^{-1} g^{-1}, respectively. The In-doped ZnS can form shallow donor whereas Cu-doping give rise to trap energy levels. The adequate addition of In and Cu into ZnS promotes a significant increase in the catalytic performance in comparison to the pristine ZnS to photoelectrochemical water splitting cells.[81]

Another alternative to increase ZnS absorption range of the solar spectrum is the combination with a narrow band gap semiconductor, such as CuS. CuS-ZnS p-n junction has been applied to photoelectrochemical water splitting. CuS thin film with hexagonal crystal structure and worm-like morphology has been obtained. The ZnS n-type semiconductor presenting hexagonal crystal structure has been deposited on CuS in order to obtain a p-n junction. The photoelectrochemical measurements performed using Na_2SO_4 solution (pH 7.5) showed a remarkable increase of photocurrent was observed reaching to 13.6 μA cm^{-2} at -0.4 V (*vs.* Ag/AgCl) in comparison to 2.6 μA cm^{-2} for pristine CuS. It was concluded that the electric field built in p-n junction promoted an electro-hole separation improvement leading to an increased photocurrent.[82]

GaN is a non-toxic material that presents a suitable position of bands but does not show satisfactory stability performance when used in photoelectrochemical cells. In this sense, the combination of nanostructured ZnS with GaN forming photoanodes in a ZnS/GaN heterostructure is an interesting arrangement to use these semiconductors. This behavior is observed due to type-II bands alignment, which promotes an improvement of the water splitting process in a photoelectrochemical cell. An optimum ZnS/GaN photoanode has shown a photocurrent density at zero bias 1.75 times higher than that of the reference GaN photoanode. Besides that, the ZnS coating on GaN promoted a better resistance to photocurrent degradation during water splitting for up 6 hours.[83]

The low photoeletrocatalytic activity of some semiconductors due to structural defects in the material, which can act as recombination centers of photogenerated charges. However, DFT (density functional theory) calculations have shown that the presence of defects can introduce new intermediate energy levels within the band gap region of the material, thus allowing the extent absorption in the visible light region; in addition, increasing the promotion of charge separation. In agreement with the calculation, the experimental results have indicated that this property can be achieved by adjusting the ZnS synthesis conditions, resulting in a band gap reduction from 3.6 to approximately 2.4 eV. This decrease was observed in the photocurrent values, in which 1.6 mA cm^{-2} was generated under visible light irradiation, in comparison to 0.25 mA cm^{-2} for the bulk region of the pure material.[84]

Semiconductor photoelectrochemical catalysis is related with factors such as band gap value, stability under illumination, generation and fast separation of electron-hole pars.

Besides, the band alignment is another crucial factor to promote the redox reactions on the photoanode and cathode. Zinc telluride has an important characteristic for the hydrogen evolution reaction, that is, its energy value of the conduction band position. The value of -1.63 V_{RHE} is the most negative conduction edge offering a large driving force for interfacial electron transfer from the semiconductor material to acceptors in the electrolyte. This semiconductor has a suitable band gap of 2.26 eV, an interesting factor for the effective light harvesting. However, the pristine ZnTe semiconductor presents some limitations, such as, slow HER kinetics related to the high overpotential and low stability to photocorrosion, which make this semiconductor non-viable to use as a photocathode in a practical device.[85]

One alternative reported in the literature to overcome this limitation was the ZnTe electrode modified with MoS_2 and carbon. One of the ZnTe-based electrodes in water splitting is the MoS_2/C/ZnTe/ZnO exhibiting an enhanced photocurrent of -1.48 mA cm^{-2} at 0 V_{RHE} in contrast with -0.19 mA cm^{-2} to the ZnTe/ZnO electrode. The reasons for the enhanced photocatalytic activity include: (a) carbon structures (fullerene and graphene), (b) the increasing stability of the photocathode, (c) MoS_2 as chemically stable electrocatalytic material for HER.[85]

CdS, CdSe and CdTe are n-type chalcogenides that exhibit an E_g of 2.2-2.4 eV, which is in accordance of a great sunlight absorption. For HER, CdS presents the conduction and the valence bands in a suitable position to promote the H^+/H_2 (0 eV) and O_2/H_2O (1.23 eV) reactions. In other words, the energy value of the conduction band is -0.9 eV whereas the value of valence band is 1.5 eV.[101]

This material presents some characteristics that hinder high performance of CdS as a photocatalyst: the lack of active sites from the easy aggregation of CdS nanoparticles added to the fast recombination of photogenerated charge carriers; and the anodic decomposition. The photocorrosion can be minimized by the presence of sacrificial reagents in the electrolyte[102] and the fast photogenerated electron-hole recombination can be mitigated by factors like preparation conditions, particle size, crystallinity and morphology of the materials.[103]

The CdS nanocomposites with Pt on graphene oxide (GO) with increase in specific surface area of CdS have revealed better results. The deposition of CdS onto GO has increased the specific surface area whereas Pt acted as the best electrocatalyst for HER.[101] Another alternative to suppress charge recombination is heterojunction formation through loaded non-noble metals on the surface of CdS.[104–106] This arrangement has to be thought in the context of the band gap and potential values of CB and VB for each material. The literature has shown that for WS_2/CdS/ITO electrode the photocurrent density has been lower than the bare CdS/ITO and CdS/WS$_2$/ITO heterojunction exhibiting a threefold stable photocurrent density in comparison to CdS/ITO electrode [87]. CdSe exhibits a direct band gap of 1.74 eV is an important material because of its exceptional optoelectronic properties in the visible range.[107]

CdSe has been used to modify the TiO$_2$ semiconductor in order to reduce the recombination charge carriers during solar light illumination. In comparison with pure TiO$_2$ nanotubes, CdSe has shown an improvement of about 10% in the photocurrent under solar illumination. This result was achieved with an optimal amount of CdSe incorporated into TiO$_2$ nanotubes by chemical bath deposition.[88] The literature has shown a band gap matched fabrication of CdSe QD/WS$_2$ nanosheets composite photocatalyst. The highest hydrogen evolution rate of 14.07 mmol h^{-1} and quantum efficiency of 58% at wavelength λ = 420 nm were achieved when the QDs size was 7-8 nm and the band gap of QDs was 1.47 eV [89]

α-Fe$_2$O$_3$ has low band gap value of 2.1 V and presents some interesting properties to PEC water splitting applications including suitable valence band and excellent chemical stability. CdSe quantum dots have been used to sensitize electrodeposited hematite thin films with low band gap material. The suitable band alignment between CdSe QDs and α-Fe$_2$O$_3$ improved the charge transfer kinetics and the absorption. Besides that, the sensitize sample exhibited a more negative flat band potential and lower resistance, resulting in a photoelectrode that achieved a photocurrent density of 1.14 mA cm^{-2} at 0.8 V vs. SCE.[90] CdSe has also been used to modified hematite photoanodes in order to obtain Ti:Fe$_2$O$_3$/CdSe heterojunction. After the CdSe modification, a photocurrent density of 3.25 mA cm^{-2} was achieved at 1.2 V vs. RHE, which is 6.5 times higher than pristine Ti:Fe$_2$O$_3$.[91] TiO$_2$ and ZnO were materials with wide band gaps, 3.0 eV and 3.37 eV, respectively, that no acts on the visible light absorption of sunlight spectrum.[108] In the context of low band gap materials combined with large band gap materials, CdS and CdSe materials have been intensively utilized to sensitize TiO$_2$ materials.[90,92–94,109] ZnO sensitized by CdSe quantum dots have been investigated as photoanode for PEC water splitting applications under visible light. The annealed CdSe (QD)/ZnO nanorods (NR) arrays films have exhibited a higher photocurrent density of 2.48 mA cm^{-2} at zero applied voltage and 100 mW cm^{-2} than that to pure ZnO NR array (as 0.41 mA cm^{-2})[95]. The sensitization of ZnO by CdTe quantum dots has also been studied. These chalcogenides have the more favorable conduction band position (E$_{CB}$ = -1.0 V vs. NHE) than CdSe (E$_{CB}$ = -0.6 V vs. NHE) and this property can improve the injection of electrons into ZnO. The combination of CdTe QDs and ZnO nanowires for photoelectrochemical water splitting has resulted in a three times higher photocurrent density, 2.0 mA cm^{-2}, than that of pristine ZnO nanowires, 0.7 mA cm^{-2}.[96] CdS/ZnO shell/core has been prepared to photocatalytic hydrogen production. To hydrothermally grow long single crystal ZnO nanoarrays, covered with CdS by chemical bath deposition, FTO has been used as substrate. With the improvement in CdS crystallinity, it has been possible to form a photocathode showing an optimized photocurrent density of 5.1 mA cm^{-2}.[97] The sensitization of wide band gap materials with low band gap quantum dots materials has also been an interesting part of research in PEC study. CdSe QDs sensitized porous hematite films has been used as photoelectrode for hydrogen generation in PEC system and the best photocurrent density achieved was 550 μA cm^{-2} at 1.0 V vs SCE. [98] CdSe nanoparticles sensitization in TiO_2 nanotubes arrays has exhibited high efficiency of hydrogen generation and effective degradation of organic pollutants in a PEC device.[110] Thin films of CdSe nanoparticles have fabricated by a gas/liquid interface reaction. The photoanodes in Na_2SO_3 solution have achieved a photocurrent density of 4.4 mA cm^{-2} at 0.69 V vs. RHE and a conversion efficiency of 4.31 % under of 100 mW cm^{-2} visible light illumination.[99]

A general overview of the binary semiconductors has been presented in this chapter, Table 1 summarizes some of the materials applied in PEC water splitting and their respective photocurrents. Besides, at the end of the chapter there is an overview of band edge positions for some binary and ternary semiconductors with respect to the vacuum level (Fig. 5).

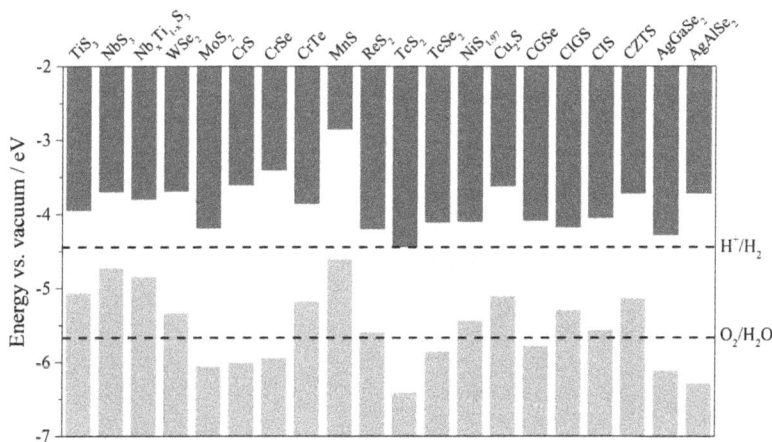

Figure 5: Band edge positions of binary and ternary semiconductors with respect to the vacuum level.

Table 1. Summary of some binary semiconductor information

Electrode	Condition	Photocurrent density	Additional Information	Ref.
TiS_3 nanoribbons	0.5 mol L^{-1} Na_2SO_3 buffered at pH 9 Halogen lamp (270 mW cm^{-2})		1.94 μmol H_2 min^{-1} at 0.3 V vs. Ag/AgCl Efficiency (η) 7.3%	[24]
$Nb_xTi_{1-x}S_3$ NbS_3	0.5 mol L^{-1} Na_2SO_3 at pH 9.5 Tungsten lamp (200 mW cm^{-2})		$Nb_xTi_{1-x}S_3$: 2.2 μmol H_2 min^{-1} cm^{-2} NbS_3: 0.65 μmol H_2 min^{-1} cm^{-2} both at 0.3 V vs. Ag/AgCl	[26]
MoS_2 Au-MoS_2 nanosheets	0.1 mol L^{-1} KH_2PO_4 at pH 7 150 W Halogen lamp with 420 nm filter (350 mW cm^{-2})	MoS_2: 100 μA cm^{-2} Au-MoS_2: 370 μA cm^{-2} (790 μA cm^{-2} after annealing) both at 0.8 V vs. Ag/AgCl		[31]
MoS_2/WSe_2	0.5 mol L^{-1} Na_2SO_4 High-power mercury lamp (100 W)	28 μA cm^{-2} at 1 V	IPCE 0.3% at 400 nm	[36]
WSe_2 nanoflakes	1.0 mol L^{-1} H_2SO_4 450 W Xe lamp with AM 1.5G filter Pt-Cu cocatalyst	Up to 4 mA cm^{-2} at 0 V vs RHE	IQE 60% at 740 nm	[42]
p-Si/ReS_2 nanosheets	0.5 mol L^{-1} H_2SO_4 Xe lamp with with AM 1.5G filter	9.0 mA cm^{-2} at 0 V vs. RHE	Half-cell STH 0.28%	[53]
FeS_2 nanoparticles	0.1 mol L^{-1} phosphate buffer at pH 7 300 W Xe lamp with a 390 nm filter	3.3 mA cm^{-2} at -0.80 V vs. NHE		[60]
FeS_2/TiO_2 nanotubes	Phosphate buffer at pH 7.4 300 W Xe lamp with AM 1.5G filter	0.917 mA cm^{-2} at 1.23 V vs. RHE		[59]
$NiS_{1.97}$ nanoparticles	0.1 mol L^{-1} Na_2SO_4 at pH 7.4 400 W Xe lamp (100 mW cm^{-2})	1.25 mA cm^{-2} at 0.68 V vs. NHE		[61]
Cu_2S Cu_2S/CdS core-sheath nanowire	0.1 mol L^{-1} KCl at pH 7 500 W Hg-Xe lamp with 330 nm filter (38 mW cm^{-2})	Cu_2S: 0.11 mA cm^{-2} Cu_2S/CdS: 0.14 mA cm^{-2} both at -0.50 V vs. Ag/AgCl		[71]
Cu_2S/CQD nanowires	1.0 mol L^{-1} KCl at pH = 5.97 500 W Xe lamp AM 1.5G	1.05 mA cm^{-2} at 0 V vs. NHE	Efficiency (η) 0.148%	[69]

$Cu_{1.94}S$ nanoflakes	0.1 mol L^{-1} KCl at pH 6.5 100 W Xe lamp with full-spectrum illumination (200 mW cm^{-2})	0.597 mA cm^{-2} at 0 V vs. RHE		[72]
CdS/Cu_2S core-shell nanorods	0.5 mol L^{-1} Na_2SO_4 at pH 7.2 and 0.2 mol L^{-1} Na_2SO_3 (hole scavenger) 150 W AM 1.5G	8.12 mA cm^{-2} at 0.5 V vs. RHE	ABPE 6.79%	[73]
Cu_2S/Cu_2O nanowires	1 mol L^{-1} Na_2SO_4 at 4.9 AM 1.5G	5.05 mA cm^{-2} at 0 V vs. RHE		[74]
Cu_2S/CdS/TiO_2/RuO$_x$	0.5 mol L^{-1} Na_2SO_4 and 0.1 mol L^{-1} KH_2PO_4 at pH 5.0 100 W Xe lamp with AM 1.5G filter	7.0 mA cm^{-2} at -0.3 V vs. RHE		[75]
graphene/CdS/Ag_2S nanofilms	0.1 mol L^{-1} Na_2S 500 W Xe lamp with AM 1.5G filter	5.18 mA cm^{-2} at 0.55 V vs. Ag/AgCl	221 µmol H_2 h^{-1} at 0.55 V vs. Ag/AgCl Efficiency (η) 4.11%	[78]
CZTS-Ag_2S nanoparticles	1.0 mol L^{-1} $Eu(NO_3)_3$ 300 W Xe lamp with 420 nm filter	0.58 mA cm^{-2} at 0.5 V vs. Ag/AgCl		[80]
4In4CuZnS nanoparticles	0.1 mol L^{-1} Na_2S at pH 3 350 W Xe lamp	12.2 mA cm^{-2} at 1.1 V vs. Ag/AgCl	1189.4 µmol h^{-1} g^{-1}	[81]
CuS-ZnS	1 mol L^{-1} Na_2SO_4 at pH 7.5 350 W Xe lamp with 420 nm filter	0.0136 mA cm^{-2} at -0.4 V vs. Ag/AgCl		[82]
ZnS/GaN	1 mol L^{-1} NaOH 300 W Xe lamp (500 mW cm^{-2})	2.8 mA cm^{-2} at 0 V vs. Ag/AgCl		[83]
ZnS nanostructured	0.24 mol L^{-1} Na_2S and 0.35 mol L^{-1} Na_2SO_3 300 W Xe lamp with 420 nm filter	1.6 mA cm^{-2} at 1.0 V vs. Ag/AgCl		[84]
MoS_2/C/ZnTe/ZnO	0.5 mol L^{-1} Na_2SO_4 AM 1.5G filter	1.48 mA cm^{-2} at 0 V vs. RHE		[85]
CdS nanorods	2 mmol L^{-1} Na_2SO_3 and 2.8 mmol L^{-1} 300 W Xe lamp with 420 nm filter (100 mW cm^{-2})	2.63 mA cm^{-2} at OCP		[86]
CdS/WS_2/ITO	0.5 mol L^{-1} Na_2S 500 W Xe lamp with 420 nm filter (55 mW cm^{-2})	0.350 mA cm^{-2} at 0.1 V vs. SCE		[87]
CdSe/TiO_2 nanotubes	1 mol L^{-1} KOH 150 W Xe lamp (100 mW cm^{-2})	1.6 mA cm^{-2} at -0.6 V vs. Ag/AgCl		[88]
CdSe QD/WS_2 nanosheets	0.5 mol L^{-1} Na_2SO_4 300 W Xe lamp with 420 nm filter		14.07 mmol h^{-1} at λ = 420 nm	[89]

CdSe QD/α-Fe$_2$O$_3$	1 mol L^{-1} NaOH (150 mW cm^{-2})	1.14 mA cm^{-2} at 0.8 V vs. SCE	[90]
Ti:Fe$_2$O$_3$/CdSe	0.25 mol L^{-1} Na$_2$S and 0.35 mol L^{-1} Na$_2$SO$_3$ at pH 13.6 300 W Xe lamp with AM 1.5G filter	3.25 mA cm^{-2} at 1.2 V vs. RHE	[91]
CdS QDs/TiO$_2$ nanotube	1 mol L^{-1} Na$_2$S AM 1.5G (138.4 mW cm^{-2})	7.82 mA cm^{-2} at -0.65 V vs. Ag/AgCl	[92]
CdSe/TiO$_2$ nanotube	0.1 mol L^{-1} Na$_2$S 500 W Xe lamp with 420 nm filter	7.1 mA cm^{-2} at -0.6 V vs. Ag/AgCl	[93]
ZnO/CdTe core-shell	0.5 mol L^{-1} Na$_2$SO$_4$ 500 W Xe lamp with 420 nm filter (100 mW cm^{-2})	4.56 mA cm^{-2} 0.8 V vs. Ag/AgCl	[94]
CdSe QD/ZnO nanorods	0.25 mol L^{-1} Na$_2$S and 0.35 mol L^{-1} Na$_2$SO$_3$ at pH 13.3 300 W Xe lamp with AM 1.5G	2.48 mA cm^{-2} at 0 V vs. SCE	[95]
CdTe QD/ZnO nanowire	0.5 mol L^{-1} Na$_2$SO$_4$ at pH 6.8 Xe lamp with AM 1.5G with 390-770 nm filter	2.0 mA cm^{-2} 1.0 V vs. Ag/AgCl	[96]
CdS/ZnO	1 mol L^{-1} Na$_2$S Xe lamp (100 mW cm^{-2})	5.1 mA cm^{-2} at 0.2 V vs. Ag/AgCl	[97]
CdSe QD/α-Fe$_2$O$_3$	1 mol L^{-1} NaOH 150 W Xe lamp (150 mW cm^{-2})	0.550 mA cm^{-2} at 1.0 V vs. SCE	[98]
CdSe	1 mol L^{-1} Na$_2$SO$_3$ at pH 8.92 500 W Xe lamp with 420 nm filter (100 mW cm^{-2})	4.4 mA cm^{-2} at 0.69 V vs. RHE	[99]

4. Transition metal-based ternary and multinary chalcogenides for photoelectrochemical water splitting

4.1 P-type copper-based chalcogenides

P-type copper based chalcogenides such as the I-III-VI$_2$ ternary compounds, CuIn(S,Se)$_2$, CuGa(S,Se)$_2$, Cu(In, Ga)(S,Se)$_2$, and the I$_2$-II-IV-VI$_2$ quaternary, such as Cu$_2$ZnSnS$_4$, are highly efficient materials for photovoltaic applications due to narrow bandgap, high extinction coefficient, suitable charge-transport properties, and band gap of 1.0 up to 2.4 eV. Because of the high performance for photovoltaics, these materials have been applied into more diverse areas of photoelectrochemistry, including water splitting. Copper-based chalcopyrite and kesterite are the most active transition metal-based chalcogenides for photoelectrochemical applications. These compounds are p-type semiconductors with intrinsic defects such as Cu vacancies, acting as photocathodes in photoelectrochemical devices. In addition, generally, the edge of the conduction band is more negative than the potential required for hydrogen reduction (H^+/H_2), and as a consequence, suitable for hydrogen evolution.

These copper-based chalcogenides have been prepared by several synthetic routes, including the single source decomposition, hydrothermal as well as solvothermal reactions, and the solid phase reactions. Besides, many deposition methods have been explored to deposit ternary and quaternary copper-based chalcogenides films. The deposition techniques reported in literature include the physical methods of vaporization, electrodeposition, spray-pyrolisis, nanocrystal solution deposition, and spin-coating, etc. Ternary chalcogenides films with 1-2 μm thick layers have been found to efficiently absorbs a wide range of solar irradiation[111].

$CuInS_2$ (CIS) is an interesting material to be applied for hydrogen evolution. This semiconductor has a band gap energy of ~1.5 eV. Photoelectrodes based on pristine CIS have shown a very low photocurrent of 80 μA cm^{-2} at 0 V_{RHE} under simulated sunlight. However, the photoelectrochemical properties of these materials have been improved by increasing grain size, which enhances the charge transfer and avoids recombination. Guijarro et al. have reported an approach to promote grain growth by incorporation of Sb, Bi and Mn during sulfurization. The coalescence of the particles would reduce the density of defects in grain boundaries in the direction to improve charge transfer and collection. The CIS photoelectrodes prepared by this technique have presented photocurrents in the range 1.7-2.0 mA cm^{-2} at 0 V_{RHE} under the conditions described above. To enhance the photoactivity of $CuInS_2$ photoelectrodes, n-type thin layers (CdS or ZnS) have been deposited on the absorber layer. It results in an improved charge separation by an introduction of a p-n junction. Moreover, the n-type overlayers would cover surface defects of the $CuInS_2$ layer film to suppress interface recombination. Particularly for CdS, its conduction band is located below that of the chalcopyrite, which facilitates the charge separation. More recently, the deposition of an additional layer of TiO_2 has been reported to be beneficial for more efficient water splitting on chalcopyrite photocathodes. TiO_2 behaves like a protective layer and does not allow a direct contact of the chalcopyrite/CdS with the electrolyte solution, in addition, its deeper valence band induces the repulsion of the holes from the surface, therefore, preventing surface recombination. Although the TiO_2 layer promotes a gain, it is not the best catalyst for water reduction. Therefore, Pt layer has been employed because of their low overpotential for HER. Guijarro et al.[112] have demonstrated the fourfold (8.0 mA cm^{-2}) increase in the photocurrent using systems using photocathodes $CuInS_2$/CdS/TiO_2/Pt in comparison to the bare $CuInS_2$ electrode. A different assembly was reported by Liu et al.[113], comprising $ITO/FeOOH/CuInS_2/Pt$, to use less toxic and more abundant element-based materials. The collaborative FeOOH and Pt effects enhanced photocurrent up to -6.79 mA cm^{-2}.

$CuInGaS_2$ (CIGS) has been very efficient among copper-based chalcopyrites for water splitting due to its tunable composition (varying Cu/Ga molar ratio), structure and band-gap (1.0-1.7 eV). Marsen et al.[114] have reported a photocurrent of 16.6 mA cm^{-2} achieved for CIGS photocathodes in 0.5 M sulfuric acid, and without considerable photodegradation during photoelectrochemical experiments. In 2010, $CuInGaSe_2$ photocathodes were applied to water splitting in aqueous solution at pH = 9.5. Yokoyama et al.[115] studied the effects of the surface modification on polycrystalline thin films for hydrogen evolution. The CIGS ternary selenide thin film was covered with Pt (Pt/CIGS), producing stable photocurrent up to 16 h under reaction and turnover number over 500. An additional layer of CdS (Pt/CdS/CIGS) improved photocurrent, a 0.3 V higher overpotential for HER, and a threefold quantum efficiency when compared to Pt/CIGS. Many other surface modifications on CIGS thin film photoelectrodes have been reported in a way to improve water splitting. Mali et al.[116] reported the application of a $CuInGa(S,Se)_2$ photoelectrode containing a layer of ZnO, CIGS/CdS/ZnO with Pt nanoparticles decorating the surface, which produced a photocurrent of -32 mA cm^{-2} at -0.7 V vs Ag|AgCl in a solution pH 9.0 under a 1.5 AM illumination. The high photocurrent was due to a type II cascade structure which facilitated the charge separation at the interface junction. Chae et al.[117] demonstrated the positive effect of ZnS passivation surface on $CuGaIn(S,Se)_2$ films. The CIGS/ZnS/Pt photocathodes exhibited highly enhanced activity (~24 mA cm^{-2} at -0.3 V RHE). However, Kim et al.[118], demonstrated a highly efficient photocathode without surface modification. Bare $CuInGa(S,Se)_2$ photocathodes were prepared by grading bandgap (top-high, middle-low, bottom-high bandgaps) exhibiting photoelectrochemical activities of 6 mA cm^{-2} at 0 V and ~22 mA cm^{-2} at -0.27 V RHE. The incorporation of Pt nanoparticles on the surface of the graded bandgap films increase the photoactivity to ~26 mA cm^{-2} at -0.16 V RHE.

$CuGaSe_2$ photocathodes have been demonstrated to be highly stable photocathodes for hydrogen evolution reaction. After the demonstration by Marsen et al[114] that $CuGaSe_2$ nanocrystalline thin film presented a photocathodic current of 16 mA cm^{-2} in acidic solution, Moriya et al.[119] reported that the PEC activity of $CuGaSe_2$ modified by deposition of a CdS layer improved and evolved hydrogen for continuously 10 days. Kim et al. [120] achieved a relationship between the photoelectrochemical activity and the Ga/Cu ratio in $CuGaSe_2$ photoelectrodes. Copper-deficient materials have presented an optimal bandgap and maximum valence band position for water splitting. However, a 3.5 Ga/Cu ratio presented the highest performance. Metal substitution has also been investigated in $CuGaSe_2$ photocathodes. Partial substitution of Cu with Ag has been reported by Zhang et al.[121]. The $Ag_xCu_{1-x}GaSe_2$ thin films were deposited on Mo by molecular beam epitaxy system, producing larger grain materials with a valence band position deeper than the $CuGaSe_2$. The $Ag_xCu_{1-x}GaSe_2/CdS/Pt$ photocathodes produced a quite stable photocurrent of 8.1 mA cm^{-2} at 0 V RHE under simulated sunlight in 0.1 M Na_2SO_4 solution (pH 9.5) for over 55 h with no decrease.

Although copper-based chalcopyrite photocathodes present high efficient photoelectrochemical activity for water splitting, scarcity of Ga and In for their application which inevitably increases fabrication cost and restricts the large scale application[122]. $CuZnSn(S,Se)_2$ kesterite is a p-type earth abundant-based chalcogenide semiconductor which has been considered a promising absorber material due to its narrow band gap (~1.5 eV)[123]. The CZTS application in PEC water splitting is very recent. One of the first reports of CZTS-based photocathode describes the chalcogenide film deposited on Mo by magnetron co-sputtering. The CZTS photoelectrode generates a cathodic photocurrent, indicating its p-type character. In this work, modifications of the surface with Pt, CdS, and TiO_2 were investigated to enhance the incident photon-to-current efficiency from 0.01%, in bare CZTS, up to 40% in the surface modified CZTS photocathode (λ = 600 nm). The conversion efficiency in H_2 production using the modified CZT reached 1.2% [124]. Sivula et al.[125] optimized and compared two different electrodeposition routs for preparing CZTS in photocathodes using protecting overlayers of CdS, TiO_2, AZO, and Pt. The CZTS film with the best performance was electrodeposited by one-step method. It was compared to films prepared by sequential deposit film method which was more time consuming and therefore it is not interesting to be applied at the industry. The one-step simultaneous element electrodeposited coating was qualitatively superior in terms of morphology, photoactivity, and especially in reproducibility of the method. The photocurrent generated by bare electrodeposited CZTS films was ~1 µA cm^{-2} at 0 V RHE. However, the photocurrent dramatically enhanced after overlayers deposition, reaching up to 1.4 mA cm^{-2} by the $Mo/CZTS/CdS/AZO/TiO_2$ at -0.2 V RHE, which can be attributed to an improved charge extraction, due to favorable interface formation with ZnO and TiO_2. To improve photoelectroactivity of CZTS for water splitting, Tay et al. [126] reported a Cd substitution into Zn, producing a $Cu_2Cd_{0.4}Zn_{0.6}SnS_4$ on a TiMo conducting layer. The Cd-substitution increased grain size, suppressed ZnS secondary phase formation, and reduced Cu_{zn} antisites. A photocurrent of 17 mA cm^{-2} at 0 V RHE was realized by $Cu_2Cd_{0.4}Zn_{0.6}SnS_4/CdS/TiMo/Pt$ photocathode.

CZTS nanocrystals have been also applied to photoelectrodes for HER. Suryawanshi et al. [127] reported the application of photocathode comprised of a TiO_2 nanorod film, containing a 5 nm Zn(S,O) passivated layer, and further introducing CZTS nanocrystals. The CZTS nanocrystals were prepared in vegetable oil, as a green solvent. This photoelectrode showed a photocurrent density ~15 mA cm^{-2} at 1.23 V vs NHE. The optimum band alignment allowed that the photogenerated electrons to quickly transferred from the CB of CZTS to Zn(S,O) then to TiO$_2$. Meanwhile the photogenerated holes on the VB of TiO$_2$ directly moved first to Zn(S,O) and then to VB of CZTS.

4.2.   Silver-based chalcogenides for water splitting

The silver-based ternary chalcopyrites have been investigated as photoelectrocatalysts for water splitting. AgGaSe$_2$ has a 1.8 eV band gap, and its band edge positions correspond to the H$_2$/H$_2$O redox potential. In addition, other silver-based chalcogenides, such as AgAlSe$_2$, and (Ag$_{0.5}$Cu$_{0.5}$)(Ga$_{0.5}$Al$_{0.5}$)Se$_2$ have valence band edges lower than that of CuGaSe$_2$, and therefore, are more suitable position for water splitting. The valence band potential for AgGaSe$_2$ is 1.3 V$_{NHE}$, which is 0.4 V deeper than the value for CuGaSe$_2$. AgGaSe films with modified surface with CdS and Pt presented photocurrent of ~7.7 mA cm^{-2} and were stable for 1 h during the experiment, but their onset potentials were lower copper-based chalcopyrite materials[128].

Electrodeposited AgInS$_2$ thin films on FTO were investigated by Wang et al.[129]. The maximum photocurrent density of the films was found to be 9.3 mA cm^{-2} with a precursor Ag-deficient chalcogenide, [Ag]/[In] = 0.8. This compound showed n-type conductivity. The photoelectrochemical properties of AgInS$_2$ and Ga-doped AgInS$_2$ deposited on FTO were studied by Tseng et al.[130]. Gallium-doping increased the band gap energy from 1.8 up to 1.9 eV, once AgGaS$_2$ has a wide band gap of 2.6 eV. Ga-doped AgInS$_2$ showed improved photoactivities, ~10 mA cm^{-2}. However, they were not very stable due to photocorrosion and dissolution of the semiconductor thin films.

Conclusions

Chalcogenides are promising candidates for photoelectrochemical water splitting. We have looked at the basic principles and the recent advances of transition metal chalcogenides to reduce or oxidize water into oxygen or hydrogen. The transition metal dichalcogenides (TMDC) have attracted extensive attention because of the tunable band gap, and high absorption coefficient. Among the aforementioned chalcogenides semiconductors, Cd-base chalcogenides, showing n-type conductivity, have been widely applied for both hydrogen and oxygen production from water. They have shown a great performance in oxygen evolution reaction, and have played an important role in most of the described systems in the literature for water splitting. When it is combined to ternary and quaternary copper-based chalcopyrites, a heterojunction is formed, and the charge extraction is enhanced. These copper-based chalcopyrites, such as CIS, CIGS, and CZT,

Photoelectrochemical Water Splitting: Materials and Applications   Materials Research Forum LLC
Materials Research Foundations **70** (2020) 1-42   https://doi.org/10.21741/9781644900734-1

have been highly efficient materials for photoelectrochemical hydrogen evolution reaction. The attractiveness of these materials lies in their well-aligned conduction band for water photoelectroreduction, high absorption coefficient, large minority-carriers diffusion length and adjustable band gap dependent on the composition. Moreover, following overlayers (TiO_2, ZnO, AZO, Pt), the charge separation and photoelectrocatalytic activity improved, and long-term stable photoelectrodes were obtained. Therefore, future efforts will focus on the interfaces to produce highly stable photoelectrodes by low-cost methods of deposition and using earth abundant and less toxic elements, to reach higher photocurrents and solar conversion efficiency into hydrogen.

References

[1]   W.B. Jensen, A Note on the Term "Chalcogen", J. Chem. Educ. 74 (1997) 1063–1064. https://doi.org/10.1021/ed074p1063

[2]   IUPAC, Nomenclature of Inorganic Chemistry, RSC Publishing, 2005

[3]   D. Shriver, M. Weller, T. Overton, J. Rourke, F. Armstrong, Inorganic Chemistry, 6th ed., W.H. Freeman & Company, New York, 2008

[4]   A. Welch, L. Baranowski, P. Zawadzki, S. Lany, C. Wolden, A. Zakutayev, $CuSbSc_2$ photovoltaic devices with 3% efficiency, Appl. Phys. Express. 8 (2015) 1–5

[5]   F. Jellinek, Transition metal chalcogenides. relationship between chemical composition, structure and physical properties crystal, React. Solids 5 (1988) 323–339

[6]   O. Stroyuk, Solar Light Harvesting with Nanocrystalline Semiconductors, Springer International Publishing, Cham, 2018. https://doi.org/10.1007/978-3-319-68879-4

[7]   A. Fujishia, K. Honda, Electrochemical photolysis of water at a semiconductor electrode, Nature 238 (1972). https://doi.org/10.1038/238037a0

[8]   H. Tributsch, Electrochemical solar cells based on layer-type transition metal compounds: Performance of electrode material, Sol. Energy Mater. 1 (1979) 257–269. https://doi.org/10.1016/0165-1633(79)90044-3

[9]   H. Tributsch, Layer-type transition metal dichalcogenides - a new class of electrodes for electrochemical solar cells, Berichte Der Bunsengesellschaft Für Phys. Chemie. 81 (1977) 361–369. https://doi.org/10.1002/bbpc.19770810403

[10]   K. Ohashi, K. Uosaki, J.O. Bockris, Cathodes for photodriven hydrogen generators: ZnTe and CdTe, Int. J. Energy Res. 1 (1977) 25–30. https://doi.org/10.1002/er.4440010104

[11]   G.K. Ahluwalia, ed., Applications of Chalcogenides: S, Se, and Te, 1st ed., Springer International Publishing, Cham, 2017. https://doi.org/10.1007/978-3-319-41190-3

[12]   M. Bouroushian, Electrochemistry of the Chalcogens, in: Electrochem. Met. Chalcogenides, 1st ed., Springer-Verlag Berlin Heidelberg, Berlin, 2010. https://doi.org/10.1007/978-3-642-03967-6_2

[13]   D. Aldakov, A. Lefrançois, P. Reiss, Ternary and quaternary metal chalcogenide nanocrystals: synthesis, properties and applications, J. Mater. Chem. C 1 (2013) 3756. https://doi.org/10.1039/c3tc30273c

[14]   M. Chhowalla, H.S. Shin, G. Eda, L. Li, K.P. Loh, H. Zhang, The chemistry of two-dimensional layered transition metal dichalcogenide nanosheets, Nat. Publ. Gr. 5 (2013) 263–275. https://doi.org/10.1038/nchem.1589

[15]   J. Ding, L. Wen, H. Li, Y. Zhang, Structure and electronic properties of graphene on ferroelectric $LiNbO_3$ surface, Phys. Lett. A. 381 (2017) 1749–1752. https://doi.org/10.1016/j.physleta.2017.03.030

[16]   X. Congxin, L. Jingbo, Recent advances in optoelectronic properties and applications of two-dimensional metal chalcogenides, J. Semicond. 37 (2016). https://doi.org/10.1088/1674-4926/37/5/051001

[17]   Q.H. Wang, K. Kalantar-Zadeh, A. Kis, J.N. Coleman, M.S. Strano, Electronics and optoelectronics of two-dimensional transition metal dichalcogenides, Nat. Publ. Gr. 7 (2012) 699–712. https://doi.org/10.1038/nnano.2012.193

[18]   J.Y. Lee, J. Shin, G. Lee, C. Lee, Two-dimensional semiconductor optoelectronics based on van der waals heterostructures, Nanomaterials 6 (2016) 1–18. https://doi.org/10.3390/nano6110193

[19]   W. Jaegermann, H. Tributsch, Interfacial properties of semiconducting transition metal chalcogenides, Prog. Surf. Sci. 29 (1988) 1–167. https://doi.org/10.1016/0079-6816(88)90015-9

[20]   T. Heine, Transition metal chalcogenides: Ultrathin inorganic materials with tunable electronic properties, Acc. Chem. Res. 48 (2015) 65–72. https://doi.org/10.1021/ar500277z

[21]   A. V. Kolobov, J. Tominaga, Two-dimensional transition-metal dichalcogenides, 1st ed., Springer International Publishing, Cham, 2016. https://doi.org/10.1007/978-3-319-31450-1

[22]   M. Pumera, Z. Sofer, A. Ambrosia, Layered transition metal dichalcogenides for electrochemical energy generation and storage, J. Mater. Chem. A 2 (2014). https://doi.org/10.1039/c4ta00652f

[23]   I.J. Ferrer, M.D. Maciá, V. Carcelén, J.R. Ares, C. Sánchez, On the photoelectrochemical properties of TiS_3 films, Energy Procedia 22 (2012) 48–52. https://doi.org/10.1016/j.egypro.2012.05.219

[24]   M. Barawi, E. Flores, I.J. Ferrer, J.R. Ares, C. Sánchez, Titanium Trisulphide (TiS_3) nanoribbons for easy Hydrogen photogeneration under visible light, J. Mater. Chem. A. 3 (2015). https://doi.org/10.1039/C5TA00192G

[25]   A. Zwick, M.A. Renucci, A. Kjekshus, Raman scattering in the IVB transition-metal trichalcogenides: ZrS_3, $ZrSe_3$, $ZrTe_3$ and $HfSe_3$, J. Phys. C Solid State Phys. 13 (1980) 5603–5614. https://doi.org/10.1088/0022-3719/13/30/023

[26]   E. Flores, J.R. Ares, C. Sánchez, I.J. Ferrer, Ternary transition titanium-niobium trisulfide as photoanode for assisted water splitting, Catal. Today. 321-322 (2019) 107-112. https://doi.org/10.1016/j.cattod.2018.01.024

[27]   J. Rijnsdorp, F. Jellinek, The crystal structure of niobium trisulfide, NbS_3, J. Solid State Chem. 25 (1978) 325–328. https://doi.org/10.1016/0022-4596(78)90118-4

[28]   H. Tributsch, Photoelectrochemistry of layer-type zirconium disulfide, J. Electrochem. Soc. 128 (1981) 1261. https://doi.org/10.1149/1.2127604

[29]   S.G. Patel, T.H. Patel, M.K. Agarwal, Photoelectrochemical study of zirconium dichalcogenides, in: world renew. Energy Congr. VI, Elsevier, (2000) 958–1961. https://doi.org/10.1016/B978-008043865-8/50414-1

[30]   F. Meng, J. Li, S.K. Cushing, M. Zhi, N. Wu, Solar hydrogen generation by nanoscale p–n junction of p -type molybdenum disulfide/ n -type nitrogen-doped reduced graphene oxide, J. Am. Chem. Soc. 135 (2013) 10286–10289. https://doi.org/10.1021/ja404851s

[31]   Z. Yin, D. Chen, M. Bosman, X. Cao, J. Chen, B. Zheng, H. Zhang, Au nanoparticle-modified MoS_2 nanosheet-based photoelectrochemical cells for water splitting, Small 10 (2014) 3537–3543. https://doi.org/10.1002/smll.201400124

[32]   Z. He, W. Que, Molybdenum disulfide nanomaterials: Structures, properties, synthesis and recent progress on hydrogen evolution reaction, Appl. Mater. Today 3 (2016) 23–56. https://doi.org/10.1016/j.apmt.2016.02.001

[33]   H. Wang, C. Li, P. Fang, Z. Zhang, J.Z. Zhang, Synthesis, properties, and optoelectronic applications of two-dimensional MoS_2 and MoS_2-based heterostructures, Chem Soc Rev. 47 (2018) 6101. https://doi.org/10.1039/c8cs00314a

[34]   X.D. Li, S.Q. Wu, Z.Z. Zhu, Band gap control and transformation of monolayer-MoS_2-based hetero-bilayers, J. Mater. Chem. C 3 (2015) 9403. https://doi.org/10.1039/c5tc01584g

[35]   S. Min, G. Lu, Sites for high efficient photocatalytic hydrogen evolution on a limited-layered MoS_2 cocatalyst confined on graphene sheets—the role of graphene, J. Phys. Chem. C. 116 (2012) 25415–25424. https://doi.org/10.1021/jp3093786

[36]   J. Xiao, Y. Zhang, H. Chen, N. Xu, S. Deng, Enhanced performance of a monolayer MoS_2/WSe_2 heterojunction as a photoelectrochemical cathode, Nano-Micro Lett. 10 (2018) 60. https://doi.org/10.1007/s40820-018-0212-6

[37]   R. Tenne, A. Wold, Passivation of recombination centers in n -WSe_2 yields high efficiency (>14%) photoelectrochemical cell, Appl. Phys. Lett. 47 (1985) 707–709. https://doi.org/10.1063/1.96066

[38]   A. Kumar, P.K. Ahluwalia, Electronic structure of transition metal dichalcogenides monolayers 1H-MX_2 (M = Mo, W; X = S, Se, Te) from ab-initio theory: New direct band gap semiconductors, Eur. Phys. J. B 85 (2012) 186. https://doi.org/10.1140/epjb/e2012-30070-x

[39]   X. Yu, M.S. Prévot, N. Guijarro, K. Sivula, Self-assembled 2D WSe_2 thin films for photoelectrochemical hydrogen production, Nat. Commun. 6 (2015) 7596. https://doi.org/10.1038/ncomms8596

[40]   H. Zhou, C. Wang, J.C. Shaw, R. Cheng, Y. Chen, X. Huang, Y. Liu, N.O. Weiss, Z. Lin, Y. Huang, X. Duan, Large area growth and electrical properties of p-Type WSe_2 atomic layers, Nano Lett. 15 (2015) 709–713. https://doi.org/10.1021/nl504256y

[41]   J.J. Devadasan, C. Sanjeeviraja, M. Jayachandran, Electrosynthesis and characterisation of n-WSe_2 thin films, Mater. Chem. Phys. 77 (2003) 397–401. https://doi.org/10.1016/S0254-0584(02)00095-0

[42]   X. Yu, N. Guijarro, M. Johnson, K. Sivula, Defect mitigation of solution-processed 2D WSe_2 nanoflakes for solar-to-hydrogen conversion, Nano Lett. 18 (2018) 215–222. https://doi.org/10.1021/acs.nanolett.7b03948

[43]   W. Chen, Y. Kawazoe, X. Shi, H. Pan, Two-dimensional pentagonal CrX (X = S, Se or Te) monolayers: antiferromagnetic semiconductors for spintronics and photocatalysts, Phys. Chem. Chem. Phys. 20 (2018) 18348–18354. https://doi.org/10.1039/C8CP02470G

[44]   S.C. Riha, A.A. Koegel, X. Meng, I.S. Kim, Y. Cao, M.J. Pellin, J.W. Elam, A.B.F. Martinson, Atomic layer deposition of MnS: Phase control and electrochemical applications, ACS Appl. Mater. Interfaces. 8 (2016) 2774–2780. https://doi.org/10.1021/acsami.5b11075

[45]   C.D. Lokhande, A. Ennaoui, P.S. Patil, M. Giersig, M. Muller, K. Diesner, H. Tributsch, Process and characterisation of chemical bath deposited manganese sulphide (MnS) thin films, Thin Solid Films 330 (1998) 70–75. https://doi.org/10.1016/S0040-6090(98)00500-8

[46]   M. Dan, Q. Zhang, S. Yu, A. Prakash, Y. Lin, Y. Zhou, Noble-metal-free MnS/In_2S_3 composite as highly efficient visible light driven photocatalyst for H2 production from H_2S, Appl. Catal. B Environ. 217 (2017) 530–539. https://doi.org/10.1016/j.apcatb.2017.06.019

[47]   X. Chen, J. Zhang, J. Zeng, Y. Shi, S. Lin, G. Huang, H. Wang, Z. Kong, J. Xi, Z. Ji, MnS coupled with ultrathin MoS_2 nanolayers as heterojunction photocatalyst for high photocatalytic and photoelectrochemical activities, J. Alloys Compd. 771 (2019) 364–372. https://doi.org/10.1016/j.jallcom.2018.08.319

[48]   X. Fang, L. Cui, T. Pu, J. Song, X. Zhang, Core-shell CdS@MnS nanorods as highly efficient photocatalysts for visible light driven hydrogen evolution, Appl. Surf. Sci. 457 (2018) 863–869. https://doi.org/10.1016/j.apsusc.2018.07.012

[49]   H. Liu, B. Xu, J. Yin, F. Miao, C. Duan, X.G. Wan, Highly efficient and ultrastable visible-light photocatalytic water splitting over ReS_2, Phys. Chem. Chem. Phys. 18 (2016) 14222–14227. https://doi.org/10.1039/C6CP01007E

[50]   Y.-L. Li, Y. Li, C. Tang, Strain engineering and photocatalytic application of single-layer ReS_2, Int. J. Hydrogen Energy. 42 (2017) 161–167. https://doi.org/10.1016/j.ijhydene.2016.11.097

[51]   S. Tongay, H. Sahin, C. Ko, A. Luce, W. Fan, K. Liu, J. Zhou, Y.-S. Huang, C.-H. Ho, J. Yan, D.F. Ogletree, S. Aloni, J. Ji, S. Li, J. Li, F.M. Peeters, J. Wu, Monolayer behaviour in bulk ReS_2 due to electronic and vibrational decoupling, Nat. Commun. 5 (2014) 3252. https://doi.org/10.1038/ncomms4252

[52]   Q. Zhao, Y. Guo, Y. Zhou, Z. Yao, Z. Ren, J. Bai, X. Xu, Band alignments and heterostructures of monolayer transition metal trichalcogenides MX_3 (M = Zr, Hf; X = S, Se) and dichalcogenides MX_2 (M = Tc, Re; X=S, Se) for solar applications, Nanoscale. 10 (2018) 3547–3555. https://doi.org/10.1039/C7NR08413G

[53]   H. Zhao, Z. Dai, X. Xu, J. Pan, J. Hu, Integrating semiconducting catalyst of ReS_2 nanosheets into p-silicon photocathode for enhanced solar water reduction, ACS Appl. Mater. Interfaces 10 (2018) 23074–23080. https://doi.org/10.1021/acsami.8b04740

[54]   X. Xu, H. Zhao, R. Wang, Z. Zhang, X. Dong, J. Pan, J. Hu, H. Zeng, Identification of few-layer ReS_2 as photo-electro integrated catalyst for hydrogen evolution, Nano Energy 48 (2018) 337–344. https://doi.org/10.1016/j.nanoen.2018.03.078

[55]   Y. Jiao, L. Zhou, F. Ma, G. Gao, L. Kou, J. Bell, S. Sanvito, A. Du, Predicting single-layer technetium dichalcogenides (TcX2 , X = S, Se) with promising applications in photovoltaics and photocatalysis, ACS Appl. Mater. Interfaces 8 (2016) 5385–5392. https://doi.org/10.1021/acsami.5b12606

[56]   D. Jasion, J.M. Barforoush, Q. Qiao, Y. Zhu, S. Ren, K.C. Leonard, Low-dimensional hyperthin FeS_2 nanostructures for efficient and stable hydrogen evolution electrocatalysis, ACS Catal. 5 (2015) 6653–6657. https://doi.org/10.1021/acscatal.5b01637

[57]   J.-Y. Zhao, J.-M. Zhang, Modulating the bandgap of the FeS_2 by O and Se doping, J. Phys. Chem. C. 121 (2017) 19334–19340. https://doi.org/10.1021/acs.jpcc.7b04568

[58]   M. Cabán-Acevedo, M.S. Faber, Y. Tan, R.J. Hamers, S. Jin, Synthesis and properties of semiconducting iron pyrite (FeS_2) nanowires, Nano Lett. 12 (2012) 1977–1982. https://doi.org/10.1021/nl2045364

[59]   Y. Xin, Z. Li, W. Wu, B. Fu, Z. Zhang, Pyrite FeS_2 sensitized TiO_2 Nanotube photoanode for boosting near-infrared light photoelectrochemical water splitting, ACS Sustain. Chem. Eng. 4 (2016) 6659–6667. https://doi.org/10.1021/acssuschemeng.6b01533

[60]   C. Guo, X. Tong, X. Guo, Solvothermal synthesis of FeS_2 nanoparticles for photoelectrochemical hydrogen generation in neutral water, Mater. Lett. 161 (2015) 220–223. https://doi.org/10.1016/j.matlet.2015.08.112

[61]   R. Bhosale, S. Kelkar, G. Parte, R. Fernandes, D. Kothari, S. Ogale, $NiS_{1.97}$: A new efficient water oxidation catalyst for photoelectrochemical hydrogen generation,

ACS Appl. Mater. Interfaces. 7 (2015) 20053–20060.
https://doi.org/10.1021/acsami.5b05077

[62]   M. Jourshabani, Z. Shariatinia, G. Achari, C.H. Langford, A. Badiei, Facile
synthesis of NiS_2 nanoparticles ingrained in a sulfur-doped carbon nitride framework
with enhanced visible light photocatalytic activity: Two functional roles of thiourea, J.
Mater. Chem. A 6 (2018) 13448–13466. https://doi.org/10.1039/C8TA03068E

[63]   D. Saraf, S. Chakraborty, A. Kshirsagar, R. Ahuja, In pursuit of bifunctional
catalytic activity in PdS_2 pseudo-monolayer through reaction coordinate mapping,
Nano Energy 49 (2018) 283–289. https://doi.org/10.1016/j.nanoen.2018.04.019

[64]   F. Grønvold, E. Røst, B. Aurivillius, N.A. Eliasson, B. Thorell, On the sulfides,
selenides, and tellurides of palladium., Acta Chem. Scand. 10 (1956) 1620–1634.
https://doi.org/10.3891/acta.chem.scand.10-1620

[65]   Y. Wang, Y. Li, Z. Chen, Not your familiar two dimensional transition metal
disulfide: structural and electronic properties of the PdS_2 monolayer, J. Mater. Chem.
C 3 (2015) 9603–9608. https://doi.org/10.1039/C5TC01345C

[66]   R. Peng, Y. Ma, B. Huang, Y. Dai, Two-dimensional Janus PtSSe for
photocatalytic water splitting under the visible or infrared light, J. Mater. Chem. A 7
(2019) 603 –610. https://doi.org/10.1039/C8TA09177C

[67]   H.L. Zhuang, R.G. Hennig, Computational search for single-layer transition-metal
dichalcogenide photocatalysts, J. Phys. Chem. C. 117 (2013) 20440–20445.
https://doi.org/10.1021/jp405808a

[68]   Q. Xu, B. Huang, Y. Zhao, Y. Yan, R. Noufi, S.-H. Wei, Crystal and electronic
structures of Cu_xS solar cell absorbers, Appl. Phys. Lett. 100 (2012) 061906.
https://doi.org/10.1063/1.3682503

[69]   M. Li, R. Zhao, Y. Su, Z. Yang, Y. Zhang, Carbon quantum dots decorated Cu_2S
nanowire arrays for enhanced photoelectrochemical performance, Nanoscale 8 (2016)
8559–8567. https://doi.org/10.1039/C5NR06908D

[70]   S.C. Riha, S. Jin, S. V. Baryshev, E. Thimsen, G.P. Wiederrecht, A.B.F.
Martinson, Stabilizing Cu_2S for photovoltaics one atomic layer at a time, ACS Appl.
Mater. Interfaces 5 (2013) 10302–10309. https://doi.org/10.1021/am403225e

[71]   S. Yang, X. Wen, W. Zhang, S. Yang, Photoelectrochemistry of pure and
core/sheath nanowire arrays of Cu_2S directly grown on copper electrodes, J.
Electrochem. Soc. 152 (2005) G220. https://doi.org/10.1149/1.1859991

Materials Research Forum LLC
https://doi.org/10.21741/9781644900734-1

[72]   K. Ren, P. Yin, Y. Zhou, X. Cao, C. Dong, L. Cui, H. Liu, X. Du, Localized defects on copper sulfide surface for enhanced plasmon resonance and water splitting, Small 13 (2017) 1700867. https://doi.org/10.1002/smll.201700867

[73]   D. Ghosh, A. Ghosh, M.Y. Ali, S. Bhattacharyya, Photoactive core–shell nanorods as bifunctional electrodes for boosting the performance of quantum dot sensitized solar cells and photoelectrochemical cells, Chem. Mater. 30 (2018) 6071–6081. https://doi.org/10.1021/acs.chemmater.8b02504

[74]   Z. Li, Z. Zhang, Tetrafunctional Cu$_2$S thin layers on Cu$_2$O nanowires for efficient photoelectrochemical water splitting, Nano Res. 11 (2018) 1530–1540. https://doi.org/10.1007/s12274-017-1769-y

[75]   Y.-X. Yu, L. Pan, M.-K. Son, M.T. Mayer, W.-D. Zhang, A. Hagfeldt, J. Luo, M. Grätzel, Solution-processed Cu$_2$S photocathodes for photoelectrochemical water splitting, ACS Energy Lett. 3 (2018) 760–766. https://doi.org/10.1021/acsenergylett.7b01326

[76]   S.I. Sadovnikov, A.I. Gusev, A.A. Rempel, Artificial silver sulfide Ag$_2$S: Crystal structure and particle size in deposited powders, Superlattices Microstruct. 83 (2015) 35–47. https://doi.org/10.1016/j.spmi.2015.03.024

[77]   V.M. Huxter, T. Mirkovic, P.S. Nair, G.D. Scholes, Demonstration of bulk semiconductor optical properties in processable Ag2S and EuS nanocrystalline systems, Adv. Mater. 20 (2008) 2439–2443. https://doi.org/10.1002/adma.200702429

[78]   B. Wang, Z. Liu, J. Han, T. Hong, J. Zhang, Y. Li, T. Cui, Hierarchical graphene/CdS/Ag$_2$S sandwiched nanofilms for photoelectrochemical water splitting, Electrochim. Acta 176 (2015) 334–343. https://doi.org/10.1016/j.electacta.2015.07.021

[79]   R. Xie, J. Su, L. Guo, Ag$_2$S/CdS nanorod-array heterojunctions for efficient photoelectrochemical water splitting, Int. J. Nanotechnol. 10 (2013) 1115. https://doi.org/10.1504/IJNT.2013.058569

[80]   A. Cabot, J. Liu, J. Arbiol, A. Genç, X. Yu, Z. Luo, M. Ibáñez, G. Zhang, Y. Zhang, A. Shavel, Cu$_2$ZnSnS$_4$ –Ag$_2$S nanoscale p–n heterostructures as sensitizers for photoelectrochemical water splitting , Langmuir 31 (2015) 10555–10561. https://doi.org/10.1021/acs.langmuir.5b02490

[81]   G.J. Lee, H.C. Chen, J.J. Wu, (In, Cu) Co-doped ZnS nanoparticles for photoelectrochemical hydrogen production, Int. J. Hydrogen Energy 44 (2018) 110–117. https://doi.org/10.1016/j.ijhydene.2018.02.112

[82]   G. Heidari, M. Rabani, B. Ramezanzadeh, Application of CuS–ZnS PN junction for photoelectrochemical water splitting, Int. J. Hydrogen Energy. 42 (2017) 9545–9552. https://doi.org/10.1016/j.ijhydene.2017.01.176

[83]   M.A. Hassan, J.H. Kang, M.A. Johar, J.S. Ha, S.W. Ryu, High-performance ZnS/GaN heterostructure photoanode for photoelectrochemical water splitting applications, Acta Mater. 146 (2018) 171–175. https://doi.org/10.1016/j.actamat.2017.12.063

[84]   F. Kurnia, Y.H. Ng, R. Amal, N. Valanoor, J.N. Hart, Defect engineering of ZnS thin films for photoelectrochemical water-splitting under visible light, Sol. Energy Mater. Sol. Cells 153 (2016) 179–185. https://doi.org/10.1016/j.solmat.2016.04.021

[85]   Y.J. Jang, J. Lee, J. Lee, J.S. Lee, Solar hydrogen production from zinc telluride photocathode modifi ed with carbon and molybdenum sulfide, ACS Appl. Mater. Interfaces 8 (2016) 7748-7755. https://doi.org/10.1021/acsami.5b07575

[86]   Y. Ma, Y. Liu, Y. Bian, A. Zhu, Y. Yang, J. Pan, Controlling shape anisotropy of hexagonal CdS for highly stable and efficient photocatalytic H2evolution and photoelectrochemical water splitting, J. Colloid Interface Sci. 518 (2018) 140–148. https://doi.org/10.1016/j.jcis.2018.02.011

[87]   M. Zirak, M. Zhao, O. Moradlou, M. Samadi, N. Sarikhani, Q. Wang, H.L. Zhang, A.Z. Moshfegh, Controlled engineering of WS2 nanosheets-CdS nanoparticle heterojunction with enhanced photoelectrochemical activity, Sol. Energy Mater. Sol. Cells 141 (2015) 260–269. https://doi.org/10.1016/j.solmat.2015.05.051

[88]   C.W. Lai, K.S. Lau, P.M. Chou, CdSe/TiO2 nanotubes for enhanced photoelectrochemical activity under solar illumination: Influence of soaking time in CdSe bath solution, Chem. Phys. Lett. 714 (2019) 6–10. https://doi.org/10.1016/j.cplett.2018.10.054

[89]   Y. Zhong, Y. Shao, F. Ma, Y. Wu, B. Huang, X. Hao, Band-gap-matched CdSe QD/WS2 nanosheet composite: Size-controlled photocatalyst for high-efficiency water splitting, Nano Energy 31 (2017) 84–89. https://doi.org/10.1016/j.nanoen.2016.11.011

[90]   A. Ikram, S. Sahai, S. Rai, S. Dass, R. Shrivastav, V.R. Satsangi, Synergistic effect of CdSe quantum dots on photoelectrochemical response of electrodeposited α-Fe2O3 films, J. Power Sources 267 (2014) 664–672. https://doi.org/10.1016/j.jpowsour.2014.05.125

[91]   J. Xiong, J. Xie, P. Yang, X. Liang, C.M. Li, Significantly improve photoelectrochemical performance of Ti:Fe2O3 with CdSe modification and surface

oxidation, Int. J. Hydrogen Energy 43 (2018) 14130–14139.
https://doi.org/10.1016/j.ijhydene.2018.05.142

[92]   Y. Xie, S.H. Yoo, C. Chen, S.O. Cho, Ag$_2$S quantum dots-sensitized TiO$_2$
nanotube array photoelectrodes, Mater. Sci. Eng. B Solid-State Mater. Adv. Technol.
177 (2012) 106–111. https://doi.org/10.1016/j.mseb.2011.09.021

[93]   J. Gan, T. Zhai, X. Lu, S. Xie, Y. Mao, Y. Tong, Facile preparation and
photoelectrochemical properties of CdSe/TiO$_2$ NTAs, Mater. Res. Bull. 47 (2012)
580–585. https://doi.org/10.1016/j.materresbull.2011.12.039

[94]   Z. Liu, X. Xie, Q. Xu, S. Guo, N. Li, Y. Chen, Y. Su, Electrochemical synthesis of
ZnO/CdTe core-shell nanotube arrays for enhanced photoelectrochemical properties,
Electrochim. Acta 98 (2013) 268–273. https://doi.org/10.1016/j.electacta.2013.03.050

[95]   S.F. Hu, R.S. Liu, W.S. Chang, C.L. Yeh, N. Chouhan, J.H. Huang, C.W. Tsai,
K.H. Chen, Array of CdSe QD-sensitized ZnO nanorods serves as photoanode for
water splitting, J. Electrochem. Soc. 157 (2010) B1430.
https://doi.org/10.1149/1.3473788

[96]   H.M. Chen, C.K. Chen, Y. Chang, C. Tsai, R. Liu, S. Hu, W. Chang, K. Chen,
Quantum dot monolayer sensitized ZnO nanowire-array photoelectrodes : True
efficiency for water splitting, Angew. Chem. Int. Ed. 49 (2010) 5966–5969.
https://doi.org/10.1002/anie.201001827

[97]   W. Wu, Y. Qiu, B. Shan, K. Feng, Y. Zhang, H. Nan, Improved photocatalytic
efficiency and stability of CdS/ZnO shell/core nanoarrays with high coverage and
enhanced interface combination, Int. J. Hydrogen Energy 42 (2016) 848–857.
https://doi.org/10.1016/j.ijhydene.2016.10.135

[98]   S. Sahai, A. Ikram, S. Rai, S. Dass, R. Shrivastav, V.R. Satsangi, CdSe quantum
dots sensitized nanoporous hematite for photoelectrochemical generation of hydrogen,
Int. J. Hydrogen Energy 39 (2014) 11860–11866.
https://doi.org/10.1016/j.ijhydene.2014.05.183

[99]   M. Guo, L. Wang, Y. Xia, W. Huang, Z. Li, Fabrication of nano-CdSe thin films
from gas/liquid interface reactions and self-assembly for photoelectrochemical
hydrogen production, Int. J. Hydrogen Energy 41 (2016) 2278–2284.
https://doi.org/10.1016/j.ijhydene.2015.12.080

[100] S.K. Saraswat, D.D. Rodene, R.B. Gupta, Recent advancements in semiconductor
materials for photoelectrochemical water splitting for hydrogen production using

visible light, Renew. Sustain. Energy Rev. 89 (2018) 228–248.
https://doi.org/10.1016/j.rser.2018.03.063

[101] P. Gao, J. Liu, S. Lee, T. Zhang, D.D. Sun, High quality graphene oxide-CdS-Pt
nanocomposites for efficient photocatalytic hydrogen evolution, J. Mater. Chem. 22
(2012) 2292–2298. https://doi.org/10.1039/c2jm15624e

[102] M. Sathish, B. Viswanathan, R.P. Viswanath, Alternate synthetic strategy for the
preparation of CdS nanoparticles and its exploitation for water splitting, Int. J.
Hydrogen Energy 31 (2006) 891–898. https://doi.org/10.1016/j.ijhydene.2005.08.002

[103] J. Yu, Y. Yu, B. Cheng, Enhanced visible-light photocatalytic H$_2$-production
performance of multi-armed CdS nanorods, RSC Adv. 2 (2012) 11829-11835.
https://doi.org/10.1039/c2ra22019a

[104] Q. Li, F. Shan, B. Sun, Y. Song, F. Wang, J. Ji, Photo-assisted electrocatalysis of
CdS/MoS$_2$ hybrid for hydrogen evolution reaction: Morphology-dependent
photoelectroactivity of p-n junction photocathode under bias potential, Int. J.
Hydrogen Energy 42 (2016) 5549–5559.
https://doi.org/10.1016/j.ijhydene.2016.05.081

[105] Y. Zang, J. Lei, Q. Hao, H. Ju, CdS/MoS$_2$ heterojunction-based
photoelectrochemical DNA biosensor via enhanced chemiluminescence excitation,
Biosens. Bioelectron. 77 (2016) 557–564. https://doi.org/10.1016/j.bios.2015.10.010

[106] J. Ji, L. Guo, Q. Li, F. Wang, Z. Li, J. Liu, Y. Jia, A bifunctional catalyst for
hydrogen evolution reaction: The interactive influences between CdS and MoS$_2$ on
photoelectrochemical activity, Int. J. Hydrogen Energy 40 (2015) 3813–3821.
https://doi.org/10.1016/j.ijhydene.2015.01.075

[107] N.G.P.S. Anjana, P.K.V. Pillai, Chemical bath deposition and characterization of
CdSe thin films for optoelectronic applications, (2010) 6653–6656.
https://doi.org/10.1007/s10853-010-4756-1

[108] X. Chen, W. Shangguan, Hydrogen production from water splitting on CdS-based
photocatalysts using solar light, Front. Energy. 7 (2013) 111–118.
https://doi.org/10.1007/s11708-012-0228-4

[109] G.S. Li, D.Q. Zhang, J.C. Yu, A new visible-light photocatalyst: CdS quantum
dots embedded mesoporous TiO$_2$, Environ. Sci. Technol. 43 (2009) 7079–7085.
https://doi.org/10.1021/es9011993

[110] W. Wang, F. Li, D. Zhang, D.Y.C.L. Leung, G. Li, Photoelectrocatalytic hydrogen
generation and simultaneous degradation of organic pollutant via CdSe/TiO$_2$ nanotube

arrays, Appl. Surf. Sci. 362 (2016) 490–497. https://doi.org/
10.1016/j.apsusc.2015.11.228

[111] S. Ikeda, T. Nakamura, S.M. Lee, T. Yagi, T. Harada, T. Minegishi, M. Matsumura, Photoreduction of water by using modified $CuInS_2$ electrodes, ChemSusChem. 4 (2011) 262–268. https://doi.org/10.1002/cssc.201000169

[112] N. Guijarro, M.S. Prevot, X. Yu, X.A. Jeanbourquin, P. Bornoz, W. Bouree, M. Johnson, F. Le Formal, K. Sivula, A bottom-up approach toward all-solution-processed high-efficiency $Cu(In,Ga)S_2$ photocathodes for solar water splitting, Adv. Energy Mater. 6 (2016) 1–13. https://doi.org/10.1002/aenm.201501949

[113] Z. Liu, X. Lu, D. Chen, Photoelectrochemical water splitting of $CuInS_2$ photocathode collaborative modified with separated catalysts based on efficient photogenerated electron-hole separation, ACS Sustain. Chem. Eng. 6 (2018) 10289–10294. https://doi.org/10.1021/acssuschemeng.8b01607

[114] B. Marsen, S. Dorn, B. Cole, R.E. Rocheleau, E.L. Miller, Copper chalcopyrite film photocathodes for direct solar-powered water splitting, MRS Proc. 974 (2006) 905–974. https://doi.org/HTTPS://DOI.ORG/ 10.1557/PROC-0974-CC09-05

[115] D. Yokoyama, T. Minegishi, K. Maeda, M. Katayama, J. Kubota, A. Yamada, M. Konagai, K. Domen, Photoelectrochemical water splitting using a $Cu(In,Ga)Se_2$ thin film, Electrochem. Commun. 12 (2010) 851–853. https://doi.org/10.1016/J.ELECOM.2010.04.004

[116] M.G. Mali, H. Yoon, B.N. Joshi, H. Park, S.S. Al-Deyab, D.C. Lim, S. Ahn, C. Nervi, S.S. Yoon, Enhanced photoelectrochemical solar water splitting using a platinum-decorated CIGS/CdS/ZnO photocathode, ACS Appl. Mater. Interfaces 7 (2015) 21619–21625. https://doi.org/10.1021/acsami.5b07267

[117] S.J. Park, O.-S. Joo, S.Y. Chae, C. Jeong, Y.J. Hwang, C.-W. Kim, S.G. Han, H. Jung, B.K. Min, Enhanced photocurrents with ZnS passivated $Cu(In,Ga)(Se,S)_2$ photocathodes synthesized using a nonvacuum process for solar water splitting , J. Am. Chem. Soc. 138 (2016) 15673–15681. https://doi.org/10.1021/jacs.6b09595

[118] B. Kim, G.-S. Park, S.Y. Chae, M.K. Kim, H.-S. Oh, Y.J. Hwang, W. Kim, B.K. Min, A highly efficient $Cu(In,Ga)(S,Se)_2$ photocathode without a hetero-materials overlayer for solar-hydrogen production, Sci. Rep. 8 (2018) 5182. https://doi.org/10.1038/s41598-018-22827-3

[119] M. Moriya, T. Minegishi, H. Kumagai, M. Katayama, J. Kubota, K. Domen, Stable hydrogen evolution from CdS-Modified $CuGaSe_2$ photoelectrode under visible-

light irradiation, J. Am. Chem. Soc. 135 (2013) 3733–3735.
https://doi.org/10.1021/ja312653y

[120] J. Kim, T. Minegishi, J. Kobota, K. Domen, Investigation of Cu-deficient copper gallium selenide thin film as a photocathode for photoelectrochemical water splitting, Jpn. J. Appl. Phys. 51 (2011) 15802. https://doi.org/10.1143/jjap.51.015802

[121] L. Zhang, T. Minegishi, J. Kubota, K. Domen, Hydrogen evolution from water using $Ag_xCu_{1-x}GaSe_2$ photocathodes under visible light, Phys. Chem. Chem. Phys. 16 (2014) 6167–6174. https://doi.org/10.1039/C3CP54590C

[122] M. Liu, J. Su, Y. Chen, S. Shen, X. Feng, Towards efficient solar-to-hydrogen conversion: Fundamentals and recent progress in copper-based chalcogenide photocathodes, Nanophotonics 5 (2016) 524–547. https://doi.org/10.1515/nanoph-2016-0027

[123] A. Fairbrother, E. García-Hemme, V. Izquierdo-Roca, X. Fontané, F.A. Pulgarín-Agudelo, O. Vigil-Galán, A. Pérez-Rodríguez, E. Saucedo, Development of a selective chemical etch to improve the conversion efficiency of Zn-rich Cu_2ZnSnS_4 solar cells, J. Am. Chem. Soc. 134 (2012) 8018–8021. https://doi.org/10.1021/ja301373e

[124] D. Yokoyama, T. Minegishi, K. Jimbo, T. Hisatomi, G. Ma, M. Katayama, J. Kubota, H. Katagiri, K. Domen, H_2 evolution from water on modified Cu_2ZnSnS_4 photoelectrode under solar light, Appl. Phys. Expres. 3 (2010) 101202. https://doi.org/10.1143/apex.3.101202

[125] L. Rovelli, S.D. Tilley, K. Sivula, Optimization and stabilization of electrodeposited Cu_2ZnSnS_4 photocathodes for solar water reduction, ACS Appl. Mater. Interfaces 5 (2013) 8018–8024. https://doi.org/10.1021/am402096r

[126] Y.F. Tay, H. Kaneko, S.Y. Chiam, S. Lie, Q. Zheng, B. Wu, S.S. Hadke, Z. Su, P.S. Bassi, D. Bishop, T.C. Sum, T. Minegishi, J. Barber, K. Domen, L.H. Wong, Solution-processed Cd-substituted CZTS photocathode for efficient solar hydrogen evolution from neutral water, Joule. 2 (2018) 537–548. https://doi.org/10.1016/j.joule.2018.01.012

[127] M. Suryawanshi, S.W. Shin, U. Ghorpade, D. Song, C.W. Hong, S.S. Han, J. Heo, S.H. Kang, J.H. Kim, A facile and green synthesis of colloidal Cu_2ZnSnS_4 nanocrystals and their application in highly efficient solar water splitting, J. Mater. Chem. A 5 (2017) 4695–4709. https://doi.org/10.1039/c7ta00257b

[128] D. Huang, C. Persson, Z. Ju, M. Dou, C. Yao, J. Guo, Investigation on AgGaSe$_2$ for water splitting from first-principles calculations, EPL 105 (2014) 37007. https://doi.org/10.1209/0295-5075/105/37007

[129] C.H. Wang, K.W. Cheng, C.J. Tseng, Photoelectrochemical properties of AgInS2 thin films prepared using electrodeposition, Sol. Energy Mater. Sol. Cells. 95 (2011) 453–461. https://doi.org/10.1016/j.solmat.2010.08.030

[130] C.J. Tseng, C.H. Wang, K.W. Cheng, Photoelectrochemical performance of gallium-doped AgInS$_2$ photoelectrodes prepared by electrodeposition process, Sol. Energy Mater. Sol. Cells 96 (2012) 33–42. https://doi.org/10.1016/j.solmat.2011.09.010

Photoelectrochemical Water Splitting: Materials and Applications   Materials Research Forum LLC
Materials Research Foundations **70** (2020) 43-57   https://doi.org/10.21741/9781644900734-2

Chapter 2

Selection of Materials and Cell Design for Photoelectrochemical Decomposition of Water

G. Keerthiga*

Chemical Engineering Department, SRM Institute of Science and Technology, India

*keerthig@srmist.edu.in

Abstract

The next commercial source of renewable energy is expected to be hydrogen. Design of intelligent innovative materials and its cell design focused on developments in photoelectrochemical (PEC) efficiency. Proper selection of materials hopes to pave the way for encompassing commercialization in the PEC water decomposition. Similarly, influence on the choice of cell design for the photoelectrochemical water decomposition will guide scientist for validating their results obtained in a single cell. This chapter is aimed to enlighten researchers on large scale reaping of hydrogen from ingeniously designed cells.

Keywords

Cell Design, Material Design, Types of Photoelectrochemical System, Heterojunction Catalyst

List of abbreviations

PV	Photovoltaic
EC	Electrolyzer
PEC	Photoelectrochemical
WE	Working electrode
RE	Reference electrode
CE	Counter electrode
STHE	Solar to hydrogen efficiency

Contents

1.   Introduction

Generation of hydrogen finds an immense potential as an alternative fuel when produced from water by inexpensive means. Photoelectrochemical technology has been widely used for water decomposition as it draws its attention from the availability of abundant renewable energy source, the sun. The perishable virtues of H_2 being a clean fuel, better than gasoline due to its high energy density [1]. Water, the cheap abundant raw material is used in PEC water decomposition and H_2 being combusted in air produces water again, rendering carbon neutral process.

The desire to pursue photocatalysis for water decomposition arises from the methodology illustration from Honda and Fujishima in the 1970s [2]. Since then, the desire to decompose water remains a dismay as its woes to its challenges in its materials design, cell design and improved efficiency for its commercialization. Large scale commercialization of PEC needs amendments to the scientific and technical challenges. Emphasis on materials technology and cell design parameters will focus on the gap in commercialization.

The major components of the PEC cell are electrodes, a photoactive anode or cathode and a counter electrode immersed in an electrolyte. The primary steps in the decomposition of water involves abstraction of light, generation of hole and electrons, reduction of H^+ to H_2 gas, and simultaneous generation of O_2 from hydroxyl OH^- ion [3]. The choice of

Photoelectrochemical Water Splitting: Materials and Applications   Materials Research Forum LLC
Materials Research Foundations **70** (2020) 43-57   https://doi.org/10.21741/9781644900734-2

photoanode or photocathode materials depends on the proper positioning of conduction and valence band to suit the reduction in hydrogen. Moreover, its complex kinetics needs huge understanding for materials design and cell design. This chapter will summarize the current amendments on materials and cell design for its accomplishments on PEC water decomposition.

2. Principle and theory of water decomposition

The redox reaction for water decomposition is discussed in equation 1 and 2 [3]. As discussed in Figure 1, in the presence of light the electrons and holes are generated, and the electron flows through the circuit and combines with H^+ ions to form H_2 gas [3].

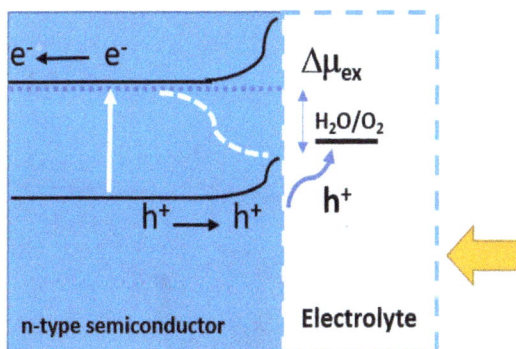

Figure 1: Representation of the photoelectrochemical water decomposition process (modified from [4])

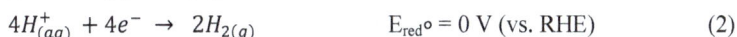

$$H_2O_{(l)} \rightarrow O_{2(g)} + 4H^+_{(aq)} + 4e^- \qquad E_{red}° = 1.23 \text{ V (vs. RHE)} \qquad (1)$$

$$4H^+_{(aq)} + 4e^- \rightarrow 2H_{2(g)} \qquad E_{red}° = 0 \text{ V (vs. RHE)} \qquad (2)$$

Basic criteria for overall water decomposition will be influenced by its position of the conduction band and valence band to suit the redox potential of H^+/H_2 and O_2/H_2O respectively [4]. The theoretically calculated potential for water decomposition is -1.23 eV (pH 7 vs RHE) with an indication of non-spontaneous Gibbs free energy of 237 kJ/mol [4]. However, resistance due to overpotentials and transportation losses leads to decomposition of water at potential near to -1.8 V [5].

The potential required to decompose water is higher than -1.23 V (vs RHE) due to major overpotential ranging from solution resistance, mass transportation limitation, kinetic limitation on the electrode surface due to light absorption and charge separation [4]. Hence, the magnificent electrodes are designed in such a way for the maximal abstraction of solar light followed by catalytic reaction.

3. Challenges in designing of a photoelectrochemical cell

Many semiconductor materials -oxides, nitrides, sulphides, carbides, and phosphides, metal oxides with d_0 and d_{10} electronic configuration and perovskite-type materials were studied for water decomposition [6]. Amidst materials technology, the proper design of a reaction cell (Figure 2, Figure 3) is needed for its technological commercialization [6]. The initiation started with trending works of fuel cell like electrochemical system on a single cell to H-type cell (Figure 2) followed by the development of the membrane electrode assembly-based process.

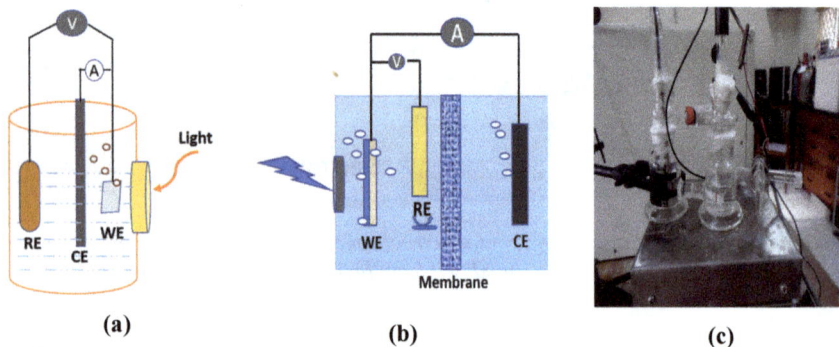

(a)   (b)   (c)

Figure 2: Pictorial representation of undivided cell (a) and divided cell (b), pictorial representation of H cell (c) for photoelectrochemical decomposition of water.

Literature suggest the use of H-type cell for separation of anode and cathode reactions and feasible separation of gaseous products from the system [7,8]. Figure 3 shows components of membrane electrode assembly for decomposition of water where continuous H_2 generation can be attempted on optimization of cell parameters.

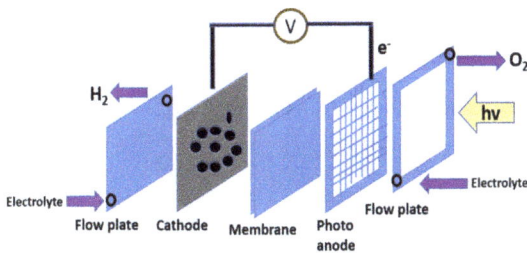

Figure 3: Illustration of membrane electrode assembly based for continuous photo electron decomposition of water

4.   Design configurations of PEC

The stabilization of intermediates followed by the evolution of H^+ as H_2 gas and improving the kinetics of rate limiting steps are the challenges involved in designing PEC cells [1]. One possible way to overcome this problem is to classify materials for the process of water decomposition into different types which are not restricted to the following [4, 9].

a) **Type I:** Mono component semiconductor materials behave as a photoactive anode or cathode exhibiting 1 D or 2 D characteristics of nanowires, nanorods, etc. (Figure 4).

b) **Type II**: A heterojunction photoelectrode device where two or more semiconductors are often coupled for several advantages. Secondary semiconductors are loaded onto the primary absorber. Cocatalyst, double and triple heterojunction are gaining attention for improved PEC (Figure 5).

c) **Type III:** The complete cell has been fabricated for both water oxidation and decomposition processes where it reports a Z scheme to perform simultaneous oxidation and reduction reactions (Figure 6).

d) **Type IV:** The resistance to transport of electrons is eliminated by use of a wireless cell where anode and cathode are fused together (Figure 7).

e) **Type V**: PV connected to electrolyser cell where a solar junction is used to reap light to generate electrons and holes which are further utilized by an electrode-electrolyte junction for water decomposition (Figure 8).

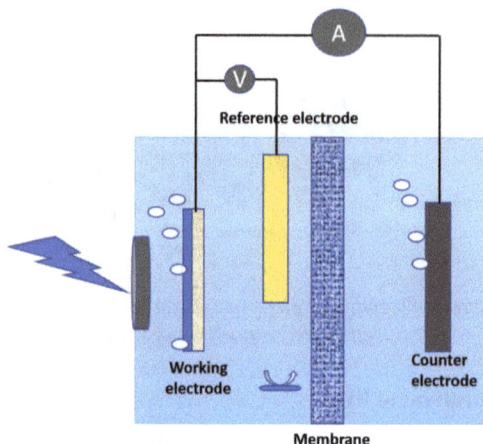

Figure 4: Type-1 photo anode type device configurations for PEC water decomposition [7]

4.1   Type 1 photo anodes

Widely used photoanodes are n-type semiconductors and in aqueous electrolyte, they perform Fermi level equalization by bending of conduction and valence band [1]. Upon illumination of light, a potential drop is created between semiconductor junctions and the cathode where for any developed potential greater than -1.23 V (vs RHE) results in decomposition of water. Amidst numerous resistances or overvoltage, the heuristic challenge in type I photoanode is to minimize the resistance for a higher incident photocurrent efficiency (IPCE) [10].

Though numerous studies on metal oxides, perovskites have been reported for Type I photoanodes (Figure 4). The obvious merits of TiO_2 are the band gap and band position for redox reaction of hydrogen evolution and stability. Tuning of material size and morphology-control, heterostructuring, quantum dots sensitizer, dopants addition, plasmonic nanoparticles, co-catalyst coupling, metal nanoparticles sensitizer, and co-catalyst coupling sensitizer have been investigated for higher IPCE [2]. They act either by increasing the conductivity, or reduced electron-hole combination, charge separation or light absorption [3]. Nontraditional way of manipulating the light-based applications is by photonic crystals, known for their light absorption characteristics and the charge separation facilitating H_2 evolution [10]. Table 1 illustrates the summary of materials for photo anode type I.

Photoelectrochemical Water Splitting: Materials and Applications   Materials Research Forum LLC
Materials Research Foundations **70** (2020) 43-57   https://doi.org/10.21741/9781644900734-2

Table 1. Summary of type I materials for water decomposition (modified from Joy et al., 2018,11)

Nanomaterial	Hydrogen formed	Reference
$SrTiO_3$ nanocrystal	276 mmol/h/gcat	12
Au multimer@MoS_2 spheres	2997.2 mmol/g	13
Pure MoS_2 spheres	881.6 mmol/g	13
Nanostructured CdS	2945 mmol/h	11
CdS nanorod/ ZnS nanoparticle	239,000 µmol/h g	11
Au–TiO_2 Nanohybrids	647,000 µmol/h g	11
$Cu_{1.94}S$-$Zn_xCd_{1-x}S$	2223 µmol/h g	11
Tantalum oxide (Ta_2O_5)	214 µmol/h g	11
g-C_3N_4 with nitrogen doped	7735 µmol/h g	13
1 wt.%Pt/C-HS-TiO_2	5713.6 µmol/h g	11

4.2   Type II heterojunction photomaterials

Type II PEC materials has been reported to exhibit a higher solar to hydrogen conversion efficiency. Amendments in Type II materials for higher H_2 yield require engineering many parameters of catalyst, electrodes, and that of the reaction conditions and the cell. Improvisation in photoanode focuses on the coupling of heterojunctions semiconductors to amplify the charge separation efficiency. The creation of double and triple layers heterojunctions (Figure 5) helps in achieving this aim by altering the band gap coherence for reduction of OH⁻ ion, reduced photo-corrosion, low cost, and stability for long hours of operation [10].

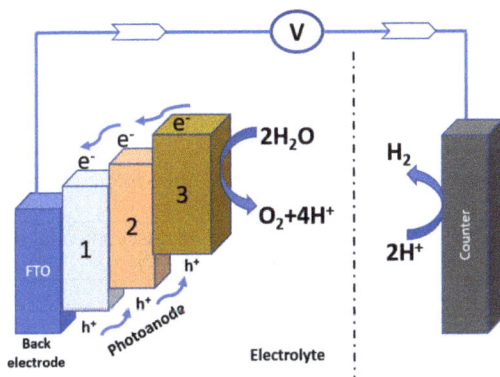

Figure 5: Schematic representation of ternary heterojunction for PEC water decomposition

Any suitable combination of semiconductors with proper band alignment should decompose water. Double heterojunction $TiO_2/BiVO_4$, $WO_3/BiVO_4$, and triple heterojunction have been used to overcome demerits of quick electron-hole combination and slow kinetics with good light absorption potential [14]. An example illustration of heterojunction formed by Ni–Mo, and Si helps to separate electrons and holes. Once electron to hole separation is achieved, the electrons can be made to reduce OH^- ion. Thus, the approach to heterojunctions is viable for charge separation in photoelectrodes. The type-II heterojunction photoanodes include $BiVO_4/WO_3$, $CaFe_2O_4/TaON$, $BiVO_4/ZnO$, and TiO_2/Fe_2O_3 [2].

Cocatalyst also works in a similar way for practical energy conversion applications which target lowered reduction potential with cost reduction when studied for H_2 evolution. Inherent advantage for the addition of cocatalyst are to decrease the overpotential, increase the kinetics of the desired reaction, to offer a metal oxide interface preferable for electron-hole separation and to prevent photo-corrosion [15]. Table 2 provides a summary of materials being successful for PEC decomposition of water.

Table 2. Summary of type II materials for photo decomposition of water

Material	Efficiency	Reference
$MoS_2/Mo_2C/g$-C_3N_4	Good quantum efficiency (41.4 % at 420 nm) 34 mmol/h/g >> than CdS	16
Bi_2WO_6 -Cu_3P	5.1 µmol/h/g	17
$BiVO_4/Fe_2TiO_5$	3 times enhancement in photocurrent, 3.2 mA/cm^2 at 1.23 V vs. RHE	15
Metal tri chalcogeni dophosphate $Ni_{0.9}Fe_{0.1}PS_3@$MXene	Lower overpotential (475 mV) and Tafel slope of 175 mV/dec	18
Co-Pi $BiVO_4$-Layer/WO_3	1.8 mA/cm^2 – 5 to12 times larger than bare WO_3 and $BiVO_4$	19
TiO_2 nanosphere-decorated $BiVO_4$	4.6 mA cm^2 at 1.23 V, three-fold enhancement over that of the bare $BiVO_4$	21
$TaON/Ta_3N_5$	With cocatalyst: negative onset potential of 350 mV, Without co catalyst: 650 mV	20

4.3 Type III wired type PEC tandem cells

An opportunity to use enzyme for enhanced water decomposition was achieved using wired tandem PEC cell. The superior scaffold of enzyme eliminates the use of sacrificial agents and the use of noble metal free catalyst works closer to the thermodynamic potential of water decomposition. Dye-sensitized TiO_2, carbon nitrides, carbon nanodots, Cd-based particles and In_2S_3 nanoparticles have already been studied for the incorporation of the enzyme [22]. The big picture of the paring of enzymes with PS II photoanode is to compliment sunlight absorption and to engineer Z scheme for commercialization. Figure 6 shows the illustration of type III cell for water decomposition.

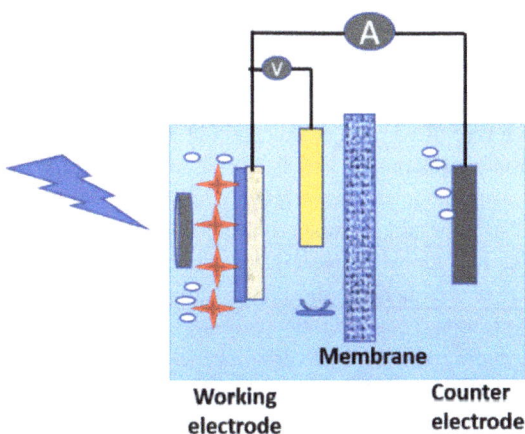

Figure 6: Schematic representation of enzyme linked cell for PEC water decomposition

Previously, NiFeSe-H_2ase and BiVO$_4$-H_2ase from *Desulfomicrobium baculatum* acted as H_2 evolution biocatalyst. The bias-free Si-based photoanode (Si|IO-TiO_2|H_2ase) uses hydrogenase and PSII for overall water decomposition [23]. The feasible demonstration of a biological system as a semi-artificial platform leads to the integration of any guest (biotic and abiotic) molecule in the PEC water decomposition system. The system comprehends the disadvantage in the biological system by appropriate light absorption by photosystems I and II for water decomposition [23].

Cohen et al., have proven organometallic clusters NiFe hydrogenase and FeFe hydrogenase to aid both proton transport and gas transport pathway [22]. *Pyrococcus furiosus* can be adsorbed and covalently linked to annular tubular TiO_2 based reactor. It showed 50% higher yield than conventionally tested electrode for H_2 evolution [24]. Few studies also demonstrate the hydrogenase immobilization on polypyrrole and clay films where its compatibility is in terms of electronic and vibrionic coupling with the enzyme and the cathode needs to be validated. Enzyme-linked process of PEC improves the light absorption of photosystems I and II and mimics semi-artificial photosynthesis. The combination of synthetic chemistry with material science and its brainstorming is needed to comprehend nature's photosynthesis [23].

4.4   Type IV wireless type PEC

Wireless photodecomposition of water involves the coupling of photoanode with cathode where the bottom of the conduction band should lie negative to the top of the valence band to ensure safe absorption and transmittance of light from one material to the other [4].

Figure 7 illustrates a type of wireless combination of electrodes for photodecomposition of water. If electron transportation loss is the issue with chosen materials, then its compatibility with photoanode can be achieved by patching anode with cathode before exposing to light. But an initial study on the assessment of solar driven water decomposition efficiency for wired (4.7%) and wireless (2.5%) showed an improved efficiency difference of 2 % under 1 sun illumination [25, 26]. Similarly, $CH_3NH_3PbI_3$ perovskite single junction solar cell, a wireless system with $CoCO_3$, Mn-oxide-TiO_2 based Si PV cell offered a solar to hydrogen efficiency (STHE) of 3.25 % for H_2 gas and was found to be active for 6 h [27]. Thus, there lies a potential for the development of an artificial leaf based on wireless photosystem in near future.

4.5   Type V PV–EC systems

The commercialization of solar panel for power generation relies on the photoelectrochemical water decomposition where a photovoltaics (PVs) cell can generate current and an appended electrolysis cell can help in decomposing water (Figure 8). Though, it has not yielded maximum solar to hydrogen conversion efficiency, but upgradation of PV-EC system has quick industrial feasibility.

Type V system helps to compact the problem of light absorption by using PV cell and the efficiency of the PV system count on upgrading solar panel efficiency. Problems like cell compartment optimization including fill factor maximization; band offset minimization finds wide scope for its improvement in Type V system [5]. It offers higher efficiency for photocatalyst and PEC based systems with higher possibilities of scale up [28], while the issues of solution resistance and mass transfer limitation in a single cell set up must be addressed.

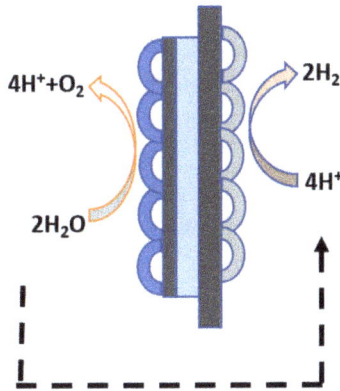

Figure 7: Schematic illustration of Type-IV wireless PEC.

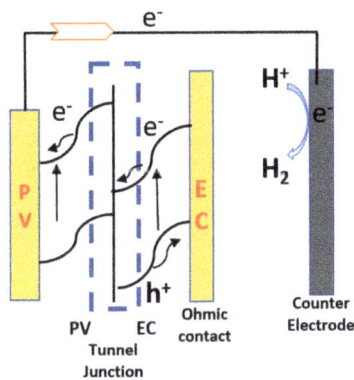

Figure 8: Schematic representation of coupled PV-EC system

Si PVs with 16% PV efficiency showed STH efficiency of 2.8% to 10% when three to four PVs were connected in series [29]. A maximum of 24.4% STH efficiency has been reported for PV EC system with high demands on parameter optimization. Jaramillo group achieved the highest ever STH efficiency (~30%) by focusing on fill factor, single to multi PV efficiency, band bending, etc. However, the general guidelines to estimate the solar to current followed by the current to hydrogen conversion does not exist in PV-EC system. Modeling can be performed with the objective of maximizing fill factor to use the system irrespective of intermediate conversion efficiency or the catalyst [28].

Thus, the commitments to new materials focus on the crystal structure, oxidation states and in general electronic configurations offering low cost and to high efficiency for water decomposition [2].

Conclusions

Type I photoelectrodes provide the choice for photo anode or photocathode as self-sustaining materials for water decomposition. Type II electrodes with heterojunction show streaming potential for new materials focusing on charge separation and hence increasing solar to hydrogen conversion efficiency. Type III systems are wired cell with tandem cell junction to provide access to biological molecules (enzymes) for light abstraction and hence improves the efficiency of the PEC system. Wireless Type IV systems focus on reduced charge transport, but their developments for commercialized system need attention. Type V PEC systems are blooming in the current solar panel era where they append the photovoltaic cell with the electrolysis cell for higher solar to hydrogen conversion efficiency. Engineering a cell with proper design idea and choice on the suitable type of PEC system will help to realize the distance dream of water decomposition in the visible region for high quantum efficiency. This chapter summarizes the cell design and materials design for a future scientist. Revolutionary development needs astounding collaborative research from scientist and technologist to visualize the dream of water powered vehicle on the road.

References

[1]   B. Viswanathan, Photo-electrochemical processes-Principles and possibilities, eprints at NCCR. https://nccr.iitm.ac.in/pecbookpart.pd. (Accessed on 15, March, 2019).

[2]   S. Cho, J. W. Jang, K.H. Lee, J. S. Lee, Strategies for efficient photoelectrochemical water splitting using metal oxide photoanodes, Appl. Mater. 010703 (2014) 2-17. https://doi.org/10.1063/1.4861798

[3]   T. Jafari, E. Moharrer, A. S. Amin, R. Miao, W. Song, S. L. Sui, Photocatalytic water splitting-the untamed dream: Review of recent advances, Mol. 21 (2016) 900-929. https://doi.org/10.3390/molecules21070900

[4]   C. Jiang, S. J. A. Moniz, A. Wang, T. Zhang, J. Tang, Photoelectrochemical devices for solar water splitting – materials and challenges, Chem. Soc. Rev. 46 (2017) 4645-4660. https://doi.org/10.1039/C6CS00306K

[5] R. Fan, S. Cheng, G. Huang, Y. Wang, Y. Zhang, S. Vanka, G. A. Botton, Z. Mi, M. Shen, Unassisted solar water splitting with 9.8% efficiency and over 100 h stability based on Si solar cells and photoelectrodes catalyzed by bifunctional Ni–Mo/Ni, J. Mater. Chem. 7 (2019) 2200-2209. https://doi.org/10.1039/C8TA10165E

[6] Z. Chen, Experimental considerations, in Z. Chen, H. N. Dinh, E. Miller (Eds), Photoelectrochemical Water Splitting, Standards, Experimental Methods, and Protocols, Springer-Verlag New York, 2013, pp.17-44. https://doi.org/10.1007/978-1-4614-8298-7_3

[7] https://nccr.iitm.ac.in/Keerthiga%20synopsisfirst%20version.docx.doc,Last accessed on 15th March, 2019.

[8] W. A. Smith, Photoelectrochemical Cell Design, Efficiency, Definitions, Standards, and Protocols, in Gimenez, J. Bisquert (eds.) Photoelectrochemical Solar Fuel Production, Springer International Publishing Switzerland, (2016) 163-197. https://doi.org/10.1007/978-3-319-29641-8_4

[9] H. W. Seo, J.S. Kim, Hydrogen Production by Photoelectrochemical Water Splitting, Appl. Sci. Converg. Technol. 27(4) (2018) 61-64. https://doi.org/10.5757/ASCT.2018.27.4.61

[10] B. S. Kalanoor, H. Seo, S. S. Kalanur, Recent developments in photoelectrochemical water-splitting using $WO_3/BiVO_4$ hetcrojunction photoanode: A review, Mater. Sci. Energy Technol. 1 (2018) 49–62. https://doi.org/10.1016/j.mset.2018.03.004

[11] J. Joy, J. Mathew, S.C. George, Nanomaterials for photoelectrochemical water splitting review, Int. J. Hydrogen Energy 43 (2018) 4804-4817. https://doi.org/10.1016/j.ijhydene.2018.01.099

[12] T. Puangpetch, T. Sreethawong, S. Yoshikawa, S. Chavadej, Hydrogen production from photocatalytic water splitting over mesoporous-assembled $SrTiO_3$ nanocrystal-based photocatalysts. J. Mol. Catal. A. Chem. 312 (2009) 97-106. https://doi.org/10.1016/j.molcata.2009.07.012

[13] X. Li, S. Guo, C. Kan, J. Zhu, T. Tong, S. Ke. Au Multimer@MoS_2 hybrid structures for efficient photo catalytical hydrogen production via strongly plasmonic coupling effect. Nano Energy. 30 (2016) 549-558. https://doi.org/10.1016/j.nanoen.2016.10.047

[14] S. Y. Tee, K. Y. Win, W. S. Teo, L.D. Koh, S. Liu, C. P. Teng, M. Y. Han, Recent progress in energy-driven water splitting, Adv. Sci. 4 (2017) 1600337-1600351. https://doi.org/10.1002/advs.201600337

[15] Y. Gao, Y. Li, G. Yang, S. Li, N. Xiao, B. Xu, S. Liu, P. Qiu, S. Hao, L. Ge, Fe_2TiO_5 as an Efficient co-catalyst to Improve the Photoelectrochemical Water Splitting Performance of $BiVO_4$, ACS Appl. Mater. Interfaces 10 (46) (2018) 39713-39722. https://doi.org/10.1021/acsami.8b14141

[16] M. Shao, Y. Shao, S. Ding, R. Tong, X. Zhong, L. Yao, W. F. Ip, B. Xu, X.Q. Shi, Yi-Y. Sun, X. Wang, H. Pan, Carbonized MoS_2: Super-active co-catalyst for high-efficient water splitting on CdS, ACS Sustainable Chem. Eng. 7(4) (2019) 4220-4229. https://doi.org/10.1021/acssuschemeng.8b05917

[17] A. Rauf, M. Ming, S. Kim, Md. S. A.S. Shah, C.H. Chung, J. H. Park, P. J. Yoo, Mediator and Co-catalysts-free direct Z-scheme composites of Bi_2WO_6-Cu_3P for solar-water splitting, Nanoscale 10 (2017) 3026-3036. https://doi.org/10.1039/C7NR07952D

[18] C.F. Du, Q. Liang, R. Dangol, J. Zhao, H. Ren, S. Madhavi, Q. Yan, Layered trichalcogenido phosphate: A new catalyst family for water splitting, Nano Micro. lett. 10(4) (2018) 67-71. https://doi.org/10.1007/s40820-018-0220-6

[19] X. Zhang, X. Wang, D. Wang, J. Ye, Conformal $BiVO_4$-layer/WO_3-nanoplate-array heterojunction photoanode modified with cobalt phosphate cocatalyst for significantly enhanced photoelectrochemical performances ACS Appl. Mater. Interfaces 11(6) (2019) 5623-5631. https://doi.org/10.1021/acsami.8b05477

[20] L. Pei, H. Wang, X. Wang, Z. Xu, S. Yan, Z. Zou, Nanostructured $TaON/Ta_3N_5$ as highly efficient type-II heterojunction photoanode for photoelectrochemical water splitting, Dalton Trans. 47 (2018) 8949-8955. https://doi.org/10.1039/C8DT01219A

[21] W. Jiang, Y. Jiang, J. Tong, Q. Zhang, S. Li, H. Tong, L. Xia, Efficient photoelectrochemical water oxidation using a TiO_2 nanosphere-decorated $BiVO_4$ heterojunction photoanode, RSC Adv. 8 (2018) 41439-41444. https://doi.org/10.1039/C8RA09072F

[22] J. Cohen, K. Kim, P. King, M. Seibert, K. Schulten, Finding gas diffusion pathways in proteins: Application to O_2 and H_2 transport in CpI [FeFe]-hydrogenase and the role of packing defects. Struct. 13 (2005) 1321–1329. https://doi.org/10.1016/j.str.2005.05.013

[23] K.P. Sokol, W. E. Robinson, J. Warnan, N. Kornienko, M. M. Nowaczyk, A. Ruff, J.Z. Zhang, E. Reisner, Bias-free photoelectrochemical water splitting with photosystem II on a dye-sensitized photoanode wired to hydrogenase, Nat. Energy 3 (2018) 944-951. https://doi.org/10.1038/s41560-018-0232-y

[24] J. Yoon, S. Bae, E. Shim, H. Joo, Pyrococcus furiosus-immobilized anodized tubular titania cathode in a hydrogen production system. J. Power Sources 189 (2009) 296–301. https://doi.org/10.1016/j.jpowsour.2008.12.072

[25] G. Kim, Ebenezer T. Igunnu George, Z. Chen, A sunlight assisted dual purpose photoelectrochemical cell for low voltage removal of heavy metals and organic pollutants in wastewater, Chem. Engg. J. 244 (2014) 411-412. https://doi.org/10.1016/j.cej.2014.01.090

[26] D. H. Nam, J. Zhang, V. Andrei, N. Kornienko, N. Heidary, A. Wagner, K. Nakanishi, K. Sokol, B. Slater, I. Zebger, S. Hofmann, J. Fontecilla-Camps, C. B. Park, E. Reisner, Solar water splitting with a hydrogenase integrated in photoelectrochemical tandem cells, Angew. Chem. Int. Ed. 57 (2018) 10595-10599. https://doi.org/10.1002/anie.201805027

[27] M. Frites, W. B. Ingler Jr., S.U.M. Khan, A single chip standalone water splitting photoelectrochemical cell, J. Tech. Innovation. Renewable Energy 314 (2014) 6-11. https://doi.org/10.6000/1929-6002.2014.03.01.2

[28] W. J. Chang, K. H. Lee, H. Ha, K. Jin, G. Kim, S.T. Hwang, H.M. Lee, S.W. Ahn, W.Yoon, H. Seo, J. S. Hong, Y. K. Go, J. I. Ha, K. T. Nam, Design Principle and Loss Engineering for Photovoltaic−Electrolysis Cell System, PV EC 2017, ACS Omega, 2 (2017) 1009−1018. https://doi.org/10.1021/acsomega.7b00012

[29] S.Y. Reece, J. A. Hamel, K. Sung, T.D. Jarvi, A. J. Esswein, J.J. H. Pijpers, D.G. Nocera, Wireless Solar Water Splitting Using Silicon-Based Semiconductors and Earth-Abundant Catalysts, Sci. 334 (2011) 645-648. https://doi.org/10.1126/science.1209816

Photoelectrochemical Water Splitting: Materials and Applications   Materials Research Forum LLC
Materials Research Foundations **70** (2020) 58-90   https://doi.org/10.21741/9781644900734-3

Chapter 3

Interfacial Layer/Overlayer Effects in Photoelectrochemical Water Splitting

Rohit Shrivastav[1*], Gurpreet Kaur[1], Divya[1], Vibha R Satsangi[2] and Sahab Dass[1]

[1]Department of Chemistry, Dayalbagh Educational Institute, Dayalbagh, Agra – 282 005, India

[2]Department of Physics & Computer Science, Dayalbagh Educational Institute, Dayalbagh, Agra – 282 005, India

*rohitshrivastav_dei@yahoo.co.in

Abstract

Aiming to gain increments in conversion efficiency, several new and innovative thoughts have been embedded in recent years to the ongoing research on photoelectrochemical splitting of water. In this context, photoelectrodes with nanoheterostructred configurations, evolved through overlayer/interfacial layer of different semiconductors, hold high promise. The approach facilitates simultaneous use of more than one low-cost and earth-abundant material in the photoelectrode design. It has emerged as an effective strategy not only to widen wavelength-range for light-absorption but also to reduce recombination losses by faster separation and swift migration of photogenerated electrons/holes. This chapter presents this specific aspect of hydrogen generation by photoelectrochemical water splitting.

Keywords

Photoelectrochemical Water Splitting, Renewable Hydrogen, Semiconductor Photoelectrode/Photocatalyst, Nanoheterostuctures, Interfacial-Layering/Over-Layering

Contents

1. Introduction

The continued world-wide interest in hydrogen production *via* photoelectrochemical (PEC) water-electrolysis and the related technology initiatives have converged to a target of roughly \$4 per kg production, which is expected to pave the way for applications in various sectors, including use in fuel cell automobiles. The PEC approach of hydrogen generation, *via* solar-water splitting, sits very high in terms of its potential for device-level developments and scalability towards mass-manufacturing [1, 2]. However, hydrogen production by this route hinges critically upon the behavior of semiconductor photoelectrode/electrolyte junction towards efficient absorption of solar light and faster separation as well as migration of photogenerated electrons/holes. There is a well-laid road-map available specifying the material characteristics desired in a semiconductor for efficient hydrogen production [3, 4]. These primarily deal with: (i) shifting the absorption of photons to higher wavelengths to augment absorption of the major part of the solar spectrum, (ii) aligning the conduction band and valence band edges of semiconductor with H_2 and O_2 evolution redox levels for spontaneous splitting of water, and (iii) eliminating/reducing photo-corrosion of the semiconductor by improving its stability. Recent years have witnessed tremendous growth in nanotechnology. It has opened up new vistas for synthesis of better materials that can be exploited as photoelectrodes for improved PEC response [5, 6]. With increase in surface area, nanostructures offer more efficient absorption of solar light. In such systems, besides the quantum confinement, local polarization and separation of photogenerated charge carriers are interesting aspects of research that have been exploited to achieve more efficient water splitting. However, any further gain in efficiency rests heavily on interfacial layer/over-layer effects generated through the use bi- and multi-layered nanoheterostructures of materials of varied bandgap, deposited one over another in layers. However, interface energetic need to be properly met in such systems for smooth transfer of photogenerated carriers, else recombination losses might prove a hurdle in the hydrogen generation. Semiconductors grown with the support of carbon-based materials, like (multi-walled carbon nanotubes (MWCNTs) and graphene, have also contributed significantly in boosting hydrogen generation. Few materials that continue to attract interest even in the midst of all above recent developments, are α-Fe_2O_3, TiO_2, ZnO and CuO [1-2, 4, 7]. Encouraging results have been obtained using above materials, particularly when prepared in nanodimension, and this has prompted many researchers to explore newer routes for their synthesis.

2.   PEC cell photoelectrode: Required characteristics and recent trends

The semiconductor photoelectrode is the key functional unit in the design of a PEC cell. Primary considerations in the identification and synthesis of semiconductor materials for application in PEC splitting of water are band structure (including band gap energy) and stability in contact with electrolyte solution (Fig. 1). Table 1 presents bandgap energy and tentative location of valence and conduction band edges of few important semiconductor materials explored over the years for PEC applications.

Table 1. Bandgap energy and tentative position of conduction and valence band edges in few important semiconductors used in PEC water splitting

Materials	Bandgap energy *(eV)*	Conduction band edge *(eV)*	Valence band edge *(eV)*	Reference
Rutile TiO_2	3.2	$3.0 - 3.7$	$-0.05 - 0.15$	[8]
AnatseTiO_2	3.2	3.2	-0.1	[8]
ZnO	3.2	$-0.25 - 0.2$	$2.95 - 3.1$	[8]
CuO	1.5	-0.51	1.04	[8]
Cu_2O	2.0	-0.7	1.3	[8]
α-Fe_2O_3	2.2	$0.3 - 0.6$	$2.4 - 2.7$	[8]
WO_3	2.6	$0.24 - 0.73$	$2.99 - 3.45$	[8]
$BiVO_4$	2.4	2.4	0.46	[8]
$SrTiO_3$	3.2	-0.6	2.6	[9]
$BaTiO_3$	3.2	0.15	3.4	[9]

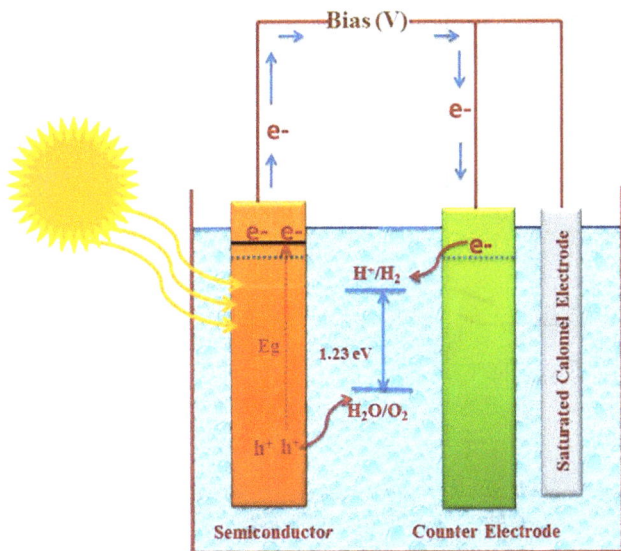

Figure 1: Schematic of a typical PEC cell used in water splitting.

The light absorption characteristics of a semiconductor are primarily guided by its band structure, with the most important being the band gap. Thermodynamically, 1.23 eV energy is needed to break water molecule into H_2 and O_2. However, considering the overpotential and other losses in an operational PEC cell circuit, the ideally required range of band gap energy for a semiconductor is 1.8-2.2 eV. Materials with very high band gap are also undesirable, as they are unable to absorb the major portion of solar radiation falling in the visible range. Another important component of band structure is the location of the band-edges, which must suitably align with H_2 and O_2 evolution redox levels (Fig. 2).

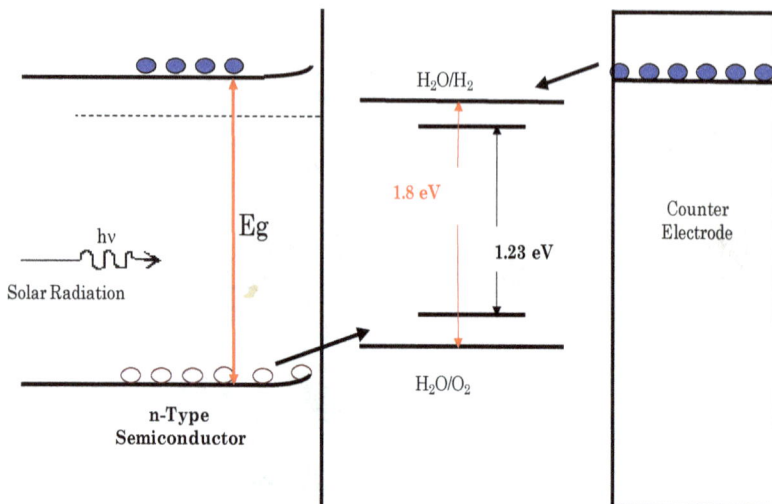

Figure 2: Ideal band-edge locations in an n-type semiconductor photoelectrode for spontaneous splitting of water.

Use of semiconductor nanoheterostructures/nanoarchitectures as photoelectrode in PEC water splitting has offered significant gain in energy conversion efficiency in recent years. With some unique optical and electrical properties and increased surface area, such material systems can serve as efficient absorber of solar light and more efficient photoelectrode in PEC cell. Further, several new and emerging ideas of material synthesis can be integrated with this research to enhance the PEC cell efficiency for water electrolysis to a level from where large-scale commercial exploitation of this technology for renewable hydrogen production becomes a reality. Use of quantum confinement, bi- and multi-layered systems, synergistic effect of low and high bandgap materials deposited in layers one over the other and multi-walled carbon nanotubes (MWCNT)/graphene supported systems are important threads of research and development in this context. Metal oxide semiconductors viz. α-Fe_2O_3, TiO_2, ZnO, $BiVO_4$, Cu_2O and CuO, which are actually old workhorses in the race, have aroused renewed interest. In various studies, nanostructures of above materials obtained through newer routes of synthesis have yielded quite encouraging PEC cell performances. The aspects of research that hold primary significance in this reference are: (a) bi- and multi-layered nanostructures of low-cost semiconductors, (b) layered nanoheterostructures supported over carbon, multi-walled carbon nanotubes and graphene and (c) Surface

Plasmon Resonance (SPR) induced enhancements in PEC performance. A combination of all above energy-harnessing and conversion strategies/mechanisms is expected to pave the way for scalability and manufacturability for large-scale hydrogen generation by PEC splitting of water.

3.   Interface layering/over-layering: An effective strategy

Sustained research of nearly five decades has revealed that while single material photoelectrodes are infected with the issues of low efficiency of energy conversion and stability in contact with electrolyte solution, nanohetrostructures prepared from suitable combinations of different materials can provide better answer to the ongoing hunt for a material system that will induce PEC water splitting spontaneously, especially under field conditions. This idea has been explored by many researchers in recent years. With a primary objective of achieving efficient light absorption and a suitable band edge energetic that would favour redox reactions at anode and cathode leading to water electrolysis, integration of different semiconductors have been studied. Several routes and protocols for material synthesis are being developed in order to obtain optimum interface layering/over-layering of different semiconductors and to explore the use of such uniquely evolved nanoheterostructrues as photoelectrode in PEC water splitting. Embedded to this new direction of research, are several other ideas of material synthesis/electrode fabrication, viz. nanostructuring/nanoarchitecture, with the involvement of carbon-based materials and metal nanoparticles. In the subsequent sections, brief discussion on these emerging lines of research along with a summary of important research reports have been presented.

4.   Interface layering/over-layering of metal oxide semiconductors

Low bandgap semiconductors are attractive from a PEC water splitting point-of-view as they can absorb visible light. However, poor transport of charge carriers imparts a serious limitation on their usage. This issue can be handled to some extent by employing different materials for the absorption of light and transport of photogenerated charge carriers. Nanoheterostructures, evolved by a suitable interface layering/over-layering of different semiconductors have been effective strategy to overcome above limitations and clear the bottleneck in solar-light induced efficient PEC water splitting [5, 10]. By suitable combinations of low and high band gap materials, photoelectrodes exhibiting better absorption of light, faster separation and flow of photoevolved electrons/holes, and greater stability, can be obtained. Depending upon the nature and band edge locations in the combining semiconductors, broadly three different kinds of band alignments at the interface of binary heterostructures can be envisaged (Fig. 3). While Type-I structure is more suitable for photocatalytic applications, the Type-II arrangement represented by Z-scheme is more favourable for PEC applications. Systems with Type-III binary junctions can also be made suitable for PEC application, albeit under external bias.

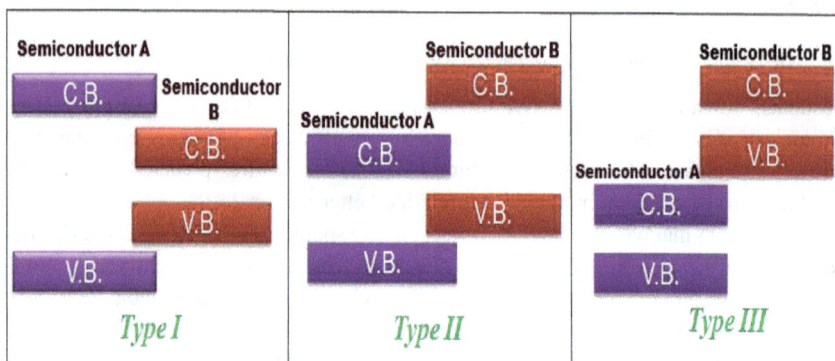

Figure 3:   Schematic representation of three different types of binary heterojunction.

4.1   Interface layering with BiVO$_4$

BiVO$_4$ is a low band gap visible light absorbing semiconductor that holds great promise for use in PEC splitting of water. However, it suffers from faster recombination of charges and very slow reaction rate for oxygen evolution on the surface. Hence, the photocatalytic activity of intrinsic BiVO$_4$ is very poor [11]. However, when it is suitably interfacially-layered with other semiconductors viz. TiO$_2$, the evolved nanoheterostructure may yield significant rise in activity. This aspect has been investigated by many researchers during the last few years and in most cases the results were highly encouraging. Given below is a brief summary of few recent research reports published on it.

Cheng et al. [11] employed fluorine-doped tin oxide (FTO) glass plates as substrate to grow BiVO$_4$/TiO$_2$ heterojunction films that were used as photoanode in PEC water splitting. They observed significant improvement in the response, compared to pristine BiVO$_4$. The effect was largely induced by the staggered configuration prevailing in BiVO$_4$/TiO$_2$ thin film heterojunction. However, the increase in photocurrent density occurred only under the bias >1.2 V vs. RHE. As in this case, the rate of hole-injection for O$_2$ evolution was slow with BiVO$_4$/TiO$_2$/FTO samples in comparison to pristine BiVO$_4$/FTO films, the bias was necessary to prevail over poor injection efficiency. PEC response was further improved with the surface-coating of cobalt phosphate co-catalyst on $BiVO_4/TiO_2/FTO$ films. In another study [12] an overlayer of $BiVO_4$ was deposited through electrodeposition, which was followed by sintering at 450°C for 2h on nanoporous TiO_2 arrays grown previously onto titanium substrate. The highest PEC current density recorded in this study was 0.3363 mA/cm^2 under the bias of 0.6 V (vs. saturated calomel electrode, SCE). Through electrochemical impedance spectroscopy it was confirmed that $BiVO_4/TiO_2/Ti$ films exhibited lowest charge transfer resistance. Resasco et al. [13] prepared $TiO_2/BiVO_4$ nanowire heterostructure photoanodes, with $BiVO_4$ working as absorber for visible light and enhanced surface area of TiO_2 facilitating charge transport. With a band alignment of Type-II in this case, electrons from $BiVO_4$ were easily transferred to TiO_2 that led to significant gain in photocurrent (2.1 mA/cm^2) under the bias of 1.23 V vs. reversible hydrogen electrode (RHE).

Interface layering of $BiVO_4$ with WO_3 has also attracted many researchers. Su et al. [14] synthesized novel heterojunction $WO_3/BiVO_4$. WO_3 nanorod array films, grown over FTO substrate, were overlayered by deposition of $BiVO_4$ through spin-coating. Interfacing with $BiVO_4$ led to faster charge separation in heterojunction and helped in reducing charge recombination, since otherwise in WO_3 alone the electron transport was regulated through diffusion. With reduced recombination and high surface area, photoconversion efficiency in this case increased significantly and it also imparted greater stability to the photoelectrode against photocorrosion. Addition of WO_3 layer beneath $BiVO_4$ photocatalyst layer was attempted by Mali et al. also [15]. They prepared thin films having nanotextured pillar morphology. When used in PEC water electrolysis, a significant improvement in response with maximum photocurrent (3.3 mA/cm^2) under 100 mW/cm^2 illuminations was recorded. The gain in photocurrent in bilayered films, relative to pristine samples of WO_3 and $BiVO_4$ was largely due to increase in surface area on deposition of $BiVO_4$ nanotextured pillars over WO_3 layer. Favorable band alignment at $WO_3/BiVO_4$ interface also helped in separation and transport of photoevolved electrons/holes.

ZnO is a high bandgap semiconductor mainly absorbing in ultraviolet region. With an objective of expanding its absorption to the visible region, attempts were made by few workers to interface-layer it with low band gap $BiVO_4$. Yan et al. [16] prepared ZnO nanorods/$BiVO_4$ heterojunction and used these as photoanode in PEC water electrolysis. ZnO nanorods were grown onto ITO substrate in a chemical bath. Subsequently, the obtained films were overlayered by $BiVO_4$ deposited through successive ionic layer adsorption and reaction (SILAR). Prepared films yielded significantly higher photocurrent (maximum 1.72 mA/cm^2) under 1.2 V applied bias (vs. Ag/AgCl), compared to pristine samples of ZnO under similar conditions. $BiVO_4$ overlayer in this case not only expanded the spectral response region but also reduced recombination losses.

Interface layering of n-type $BiVO_4$ with p-type Co_3O_4 was studied by Long et al. [17]. They prepared $BiVO_4/Co_3O_4$ powder electrodes, having 0.8 wt. % cobalt, along with pristine samples. They observed nearly 4 times gain in IPCE values in case of $BiVO_4/Co_3O_4$, compared to pristine samples. The formed n/p junction in $BiVO_4/Co_3O_4$ samples favoured separation of photogenerated charges. As a result, stabilized photocurrent, increased efficiency of energy conversion, and the desired segregation at the interface of oxidation and reduction were achieved. Forming a p/n junction, Soltani et al. [18] attempted to interface $BiVO_4$ with $BiFeO_3$ to obtain p-type $BiFeO_3$/n-type $BiVO_4$ nanoheterostructure photoelctrodes, through a facile ultrasonic/hydrothermal route. Such electrodes yielded improved efficiency for PEC water splitting in contrast with the pristine $BiVO_4$. Additionally the nanoheterostructures were stable compared to the poor stability of pristine $BiVO_4$. Interfacing of $BiVO_4$ with MoO_3 has also been investigated. Chen et al. [19] explored the use of $MoO_3/BiVO_4$ heterojunction films for PEC water splitting. They also introduced oxygen vacancies in the structure through post-annealing in argon environment that improved the PEC response even further, as photocurrent density increased to the maximum 4.1 mA/cm^2, representing nearly 200% gain over the performance of normal $MoO_3/BiVO_4$ films. Faster charge transfer and improved kinetics of water oxidation at the electrode surface primarily helped in more efficient water splitting in this case.

4.2   Interface layering with CuO/Cu_2O

Interfacing of copper oxides, viz. CuO and Cu_2O, with other metal oxides, particularly of high optical band gap, is interesting from several counts. It would not only expand the light absorption range to cover almost all wavelengths from UV to visible, but may also facilitate more efficient separation of photogenerated charge carriers. Copper oxides have been widely reported to suffer with instability. Hence, over-layering of copper oxides with materials like ZnO and TiO_2 is also expected to provide stability to the former. Shaislamov et al. [20] prepared stable films, represented by CuO/ZnO nanoheterostrcture, for use as photocathode in PEC water electrolysis. Authors first grew CuO nanorods by oxidizing Cu. Over this layer of CuO, ZnO nanobranches were then deposited by hydrothermal process. Thus obtained films when used as photoelectrode showed greater stability as ZnO layer protected CuO layer from electrolytic corrosion. However, due to electrochemical resistance in charge transfer along with minor hindrance in light absorption by CuO, PEC response was marginally reduced. In a reverse pattern of interlayering, Zhang et al. [21] first grew ZnO nanorods on Indium Tin Oxide (ITO) glass sheets by galvenostatic cathodic reduction and it was overlayered by CuO nanocrystals deposited through thermal evaporation-deposition method using low-cost copper foam as Cu source. Bilayered films yielded significant gain in PEC response with maximum efficiency of light-to-hydrogen generation being 0.86% at 0.94 V bias. Kwon et al. [22] developed thin film nanostructures (hierarchical), represented by ZnO/CuO/Au and ZnO/CuO/Ag, which were used as photoelectrode in PEC water electrolysis. ZnO was grown as an array of nanowires over Fluorine-doped Tin Oxide (FTO) glass substrate. Subsequently, it was overlayered with CuO nanostructures and Ag/Au nanoparticles. The samples yielded enhanced optical and PEC responses. Films with ZnO/CuO/Au configuration showed two-fold gain in the photocurrent compared to films with ZnO/CuO/Ag configuration. Strong surface Plasmon of Au also helped in improving the PEC response of samples. Kargar et al. [23] reported synthesis of ZnO/CuO films, represented by a three-dimensional branched nanowire heterojunction. Prepared films were used in a neutral medium for PEC water electrolysis. They observed good correlation among electrode current, size and density. Films with dense and long growth of CuO nanowires exhibited enhanced surface area that ultimately led to increase in photocurrent. The effects of interface stoichiometry in modifying the alignment of band edges in ZnO/Cu_2O heterojunctions and the device-voltage were investigated by Wilson et al. [24]. They highlighted the significance of stoichiometric interfaces and the need of reproducible experimental methods for their generation. Wu et al. [25] obtained three-dimensional nanostructures, comprising of backbone of CuO nanocones over which branches of ZnO nanorod were grown. Samples exhibited significantly improved PEC performance. Synergistic effects on separation of charge carriers with increase in their lifetime were cited as primarily responsible for gain in PEC response. Energy band alignment plays very significant role in the functioning of semiconductor devices. Hence, the challenge lies in having suitable band structure at the heterojunction. Emphasizing this aspect, Siol et al. [26] synthesized Cu_2O/ZnO heterointerfaces and observed that alteration in film preparation conditions has dramatic influence on the behaviour of the interface. Izaki et al, [27] prepared polycrystalline $n-ZnO/p-Cu_2O$ heterojunction films by electrodeposition in aqueous medium. The samples showed much improved electrical rectification with PV conversion efficiency of 1.28% under AM 1.5. Tsege et al, [28] fabricated CuO/ZnO nanowire based photocathode by an easily up-scalable method. Samples were used in PEC water electrolysis and yielded current as high as 8.1 mA/cm^2 under zero bias (vs. RHE). The increase in PEC response was primarily attributed to gain in light absorption and efficient separation of photogenerated charge carriers. CuO/ZnO interlayer p-n nano-heterojunction films were investigated by several other workers [29-30] in recent past, suggesting significant gain in PEC photocurrent for water splitting, clearly indicative of efficient charge separation and reduced recombination.

Zhang et al. [31] obtained, by hydrothermal reaction route, CuO@TiO$_2$ films composed of hierarchical nanowires. Expanded absorption of light and faster rate of charge transfer at the interface facilitated PEC water splitting. Heterostructure films yielded almost five times gain in photocurrent density against the pristine TiO$_2$ films. Atabaev et al. [32] prepared bilayered TiO$_2$/CuO films by spin coating and pulsed laser deposition and investigated use in PEC water electrolysis. Significant enhancement in photocurrent, (0.022 mA/cm^2) under the applied voltage of 1.23 V (vs. RHE) against 0.013 mA/cm^2 recorded with bare TiO$_2$, was evidenced through efficient separation of charge carriers and better absorption of solar light. In another study of interfacial layering with copper oxide, Li et al. [33] adopted 2-step anodization followed by sintering to first grow TiO$_2$ film, represented by arrays of nanotube. Subsequently, Cu nanoparticles were deposited by impregnation-reduction method to generate heterojunction, Cu$_2$O/Cu/TiO$_2$. The presence of Cu$_2$O, along with surface Plasmonic effect of copper metal, caused significant expansion in absorption wavelength from 400 to 700 nm. This resulted in stable and more efficient H$_2$ evolution, particularly on absorption of visible light. Mor et al. [34] obtained Cu-Ti-O films, represented by vertically oriented array of nanotubes. In this study, films with 1μm thickness yielded nearly 11% quantum efficiency, with complete visible spectrum contributing in PEC photocurrent generation. Siripala et al. [35] deposited n-type TiO$_2$ over previously grown p-type Cu$_2$O surface on Ti foil. Under the applied potential of 1V and 700 W/m^2 illuminations, the maximum photocurrent recorded with the prepared samples was 0.7 mA/cm^2. More importantly, TiO$_2$ overlayer film effectively protected Cu$_2$O layer from electrolytic corrosion. Mendez-Medrano et al. [36] grew on commercially procured TiO$_2$ the nanoparticles of CuO and Ag. It increased significantly the photocatalytic activity of samples, which was largely mediated through surface modification induced more efficient charge carrier separation. Surface Plasmon effect of Ag nanoparticles and the narrow band gap of CuO effectively combined to yield better absorption of visible wavelengths in this case. In another study, Moniz and Tang [37] used commercially available TiO$_2$ (P25) and an inexpensive precursor salt of copper to develop a rapid solvothermal microwave procedure to evolve CuO/TiO$_2$ heterostructure. With an intimate contact between CuO and TiO$_2$ at the junction, PEC water splitting was enhanced to almost double. Faster transfer of photoelectrons to CuO from TiO$_2$ conduction band enabled efficient participation of leftover holes in TiO$_2$ in water oxidation. Sharma et al. [38] attempted to evolve interface layering of Cu$_2$O with BaTiO$_3$ to obtain BaTiO$_3$/Cu$_2$O heterojunction thin films. Such films, on being used as phtoelectrode in PEC water splitting, yielded major gain in activity. In this work, a comparison cum validation of experimental findings through first-principles DFT computations was also successfully attempted. In an interesting work, Samarasekara [39] prepared nanoheterojunction thin films, p-type Cu$_2$O/n-type CuO, which were explored

Photoelectrochemical Water Splitting: Materials and Applications   Materials Research Forum LLC
Materials Research Foundations **70** (2020) 58-90   https://doi.org/10.21741/9781644900734-3

for optical and electrical properties. The fabrication of such unique nanoheterojunction was made possible through controlled oxidation of copper films by regulating the sintering environment. Clear diode characteristics, favouring PEC water splitting, were observed.

4.3   Interface layering with hematite (α-Fe$_2$O$_3$)

Verma et al. [40] prepared thin film heterojunction electrodes by interfacial layering of Ti-doped Fe$_2$O$_3$ with MWCNT and Cu$_2$O that yielded much-improved response for PEC water electrolysis. The maximum photocurrent recorded under 1 V applied voltage (vs. SCE) was 5.17 mA/cm^2. The gain in photocurrent was mainly attributed to increase in conductance and transport of photogenerated charge carriers. To this effect the role of MWCNT was very crucial. Investigations on thin layered heterojunction formed between α-Fe$_2$O$_3$ and Cu$_2$O was also attempted by Juodkazyte et al. [41]. They proposed that the observed effective charge separation under illumination was due to smooth transfer of charge carriers that followed Z scheme, involving faster quenching of photogenerated holes in Cu$_2$O by CB electrons of hematite, as the later travelled swiftly via interface. Sharma et al. [42] synthesized nanostructured Ti-doped α-Fe$_2$O$_3$/Cu$_2$O heterojunction photoelectrode. Such electrodes exhibited extension in the absorption wavelength range and reduced recombination that led to efficient hydrogen production. Interface layering of copper oxides with SrTiO$_3$ has also been explored. Sharma et al. [43-44] investigated PEC water splitting by employing Cu$_2$O/SrTiO$_3$ nanoheterojunction thin films as photoelectrode. Further in a separate study, Choudhary et al. [6] prepared CuO/SrTiO$_3$ nanohetrostructured bilayered thin films for use in PEC water splitting. Marked increase in photocurrent was recorded during above studies that highlighted the importance of such heterostructures in increasing the efficiency of hydrogen generation via PEC water splitting.   In another study, [45] α-Fe$_2$O$_3$ wrapped CuFeO$_2$ heterolayer structure was obtained on ITO by sol-gel and electrophoretic deposition that yielded significant gain in PEC photocurrent for water splitting. Decoration of the film surface by platinum nanoparticles, grown through photoreduction, further improved the PEC response.

Interfacial layering of α-Fe$_2$O$_3$ with NiO, to obtain n-p junction diode films, has also been attempted by few workers. Rajendran et al. [46] attempted to deposit p-type NiO nanoparticles on the previously grown n-type α-Fe$_2$O$_3$ films. They observed enhanced PEC photocurrent (highest 1.55 mA/cm^2) at 1 V applied potential (vs. RHE), against the value of 0.08 mA/cm^2 recorded with pristine samples of hematite, under similar conditions using a solar simulator for illumination. Singh and Sarkar [47] reported the synthesis of p-n heterojunction nanowire-based photoanode, represented by densely packed individual p-NiO/n-Fe$_2$O$_3$ heterojunction. Prepared by consecutive electrodeposition of Fe and Ni nanowires on anodic alumina template followed by controlled oxidation, the samples exhibited nearly 24 times gain in photocurrent at applied bias of 1.23 V (vs. RHE) against the pristine samples of hematite. They opined that the significant gain in the efficiency of PEC water splitting was due to the synergistic combination of α-Fe$_2$O$_3$ (n-type semiconductor) with NiO (p-pype semiconductor) that had facilitated generation and transfer of charge carriers. By suitably combining hematite with NiO and graphene oxide (reduced form), nanostructured photoanodes were synthesized by Phuan et al. that yielded improved PEC performances [48-49]. In this study, intimate contact between α-Fe$_2$O$_3$ and NiO, mediated through eRGO, enabled faster charge transfer and extension of spectral absorption range. Compared to bare α-Fe$_2$O$_3$, films with combination of eRGO and NiO yielded nearly three-fold gain in H$_2$ evolution rate, with maximum hydrogen generation rate 92 μmol h^{-1} cm^{-2}.

Interfacing of moderate band gap material, α-Fe$_2$O$_3$, with high bandgap semiconductors like, TiO$_2$, ZnO and SnO$_2$ can yield broader absorption of light. Exploring this possibility, Mohapatra et al. [50] synthesized hematite nanostructures on TiO$_2$ nanotubular template by pulsed electrodeposition technique. They observed that an important variant in this process was the deposition time, as by varying the time of deposition the growth of α-Fe$_2$O$_3$ can be regulated, in the range 50-550 nm, inside the nanotubes of TiO$_2$. Niu et al. [51] synthesized through epitaxial growth branched SnO$_2$ on α-Fe$_2$O$_3$ nanospindles and nanocubes. The obtained nanoheterostructures exhibited excellent photocatalytic abilities both under visible and UV irradiances, largely accredited to efficient separation of photoevolved electrons/holes. Chen et al. [52] synthesized α-Fe$_2$O$_3$/ZnO heterojunction films by hydrothermal route and used these for PEC water splitting. With ZnO overlayer, transfer of photo-evolved electrons from ZnO to hematite was facilitated, leading to gain in photocurrent. However, the thickness of ZnO affected photocurrent density. Bouhjar et al. [53] also attempted to fabricate ZnO/α-Fe$_2$O$_3$ heterojunction device for hydrogen production. They observed increase in intrinsic surface states or defects at ZnO/α-Fe$_2$O$_3$ interface. Owing to the high active interfacial surface area, higher PEC activity was recorded with bilayered films compared to pristine α-Fe$_2$O$_3$. Miao et al. [54] synthesized Ti-doped hematite films which were then overlayred with ZnFe$_2$O$_4$ to evolve heterojunction films. Prepared samples were tested for PEC water electrolysis. Ti doping in α-Fe$_2$O$_3$ improved photoelectric conversion efficiency. Further in heterojunction film samples, electron/hole separation was facilitated at interface. To assist the charge separation, there existed a favourable alignment in the band edges of ZnFe$_2$O$_4$ with Ti-doped α-Fe$_2$O$_3$. In another work, Ahmed et al. [55] prepared nanostructured p-type CaFe$_2$O$_4$/n-type α-Fe$_2$O$_3$ interlayer films on FTO. Interlayer photoanodes showed nearly 100% increase in PEC response compared to the bare α-Fe_2O_3, when the samples were subjected to 1-sun illumination (illumination intensity 100 mW/cm^2). Increase in PEC response resulted primarily from faster separation of charge carriers and fall in electrical resistance. Further overlayering of samples with CoPi co-catalyst raised the PEC response even higher.

4.4   Interface layering with WO_3

WO_3 is a semiconductor of high technological significance. It is an effective photocatalyst to oxidize water to O_2. However, WO_3 alone cannot drive overall water splitting process for the reason that the reducing power generated at its conduction band edge is not sufficient enough to induce hydrogen evolution reaction [56]. Although, quite similar to α-Fe_2O_3, WO_3 is also absorbing to the visible light, yet the conduction band edge of d orbital in this case is shifted downward, which does not support hydrogen evolution reduction reaction. With respect to the position of valence band edge corresponding to oxidizing power, WO_3 resembles closely with TiO_2. An appropriate interface layering of WO_3 with other promising semiconductors, with an objective of raising its performance towards PEC water electrolysis, has attracted several researchers in recent past. Muller et al. [57] obtained a unique nanoarchitecture by combining α-Fe_2O_3 with WO_3. It was achieved through suitable interface layering of WO_3 with α-Fe_2O_3. On previously grown Fe_2O_3 films, overlayering of WO_3 was attempted as surface treatment and also as scaffold. Heterojunction samples, when used in PEC cell, revealed improved charge separation. Wu et al. attempted to combine NiO with WO_3 [58]. In their study they obtained a novel heterojunction by overlayering of NiO with WO_3. Prepared films were used as photocathode and their performance was evaluated for PEC water electrolysis. Bilayered films offered nearly five-fold gain in PEC photocurrent than pristine NiO under simulated sunlight, largely attributable to better absorption of light coupled with efficient separation of photoevloved electrons/holes and their swift migration. In several studies WO_3 has been combined with MWO_4 (M = Zn, Ni, Co, Cu). Leonard et al. [59] obtained $ZnWO_4$/WO_3 composite. Samples with 9 % Zn/W ratio yielded nearly 150% gain in photocurrent against the pristine samples of WO_3. Band diagram in this case clearly suggests that there exists greater possibility of separation of photogenerated charge carriers in composite than in pristine WO_3. Anis et al. [60] prepared WO_3/$NiWO_4$ composite nanofibers for PEC water splitting. Maximum gain in photocurrent with composite nanostructure was nearly 70 % against the pristine WO_3 fibers under similar conditions. Here also a significant fall in the transfer resistance for charges in composites primarily contributed to observed enhancement in PEC photocurrent. Zhang et al. [61] obtained two-dimensional bilayer nanosheets of WO_3@$CoWO_4$. Prepared samples proved to be efficient catalyst for enhanced PEC reactions pertaining to the evolution of oxygen. With the unique heterostructure in

$WO_3@CoWO_4$ films, the energy barriers for oxygen evolution were diminished. Besides, more effective visible light absorption, faster transfer of charges and reduced recombination induced major improvement in PEC performance. Martinez-Garcia et al. [62] attempted the deposition of crystalline $CuWO_4$ and WO_3 on copper oxide nanowire arrays, which yielded five-fold gain in photocurrent density on TiO_2 coated copper oxide. Interfacial layer of WO_3 or $CuWO_4$ reduced CuO phase impurity in Cu_2O, which led to considerable improvement in photoactivity. Protective coatings in this work probably served the twin role. On one hand it extended the stability of electrode and on the other it facilitated the separation of charge carriers. In an interesting study, Weng et al. [63] attempted interface-layering of WO_3 with barium bismuth niobate. However, pristine samples of barium bismuth niobate (BBNO), prepared as thin film electrode, exhibited low photocurrent for PEC water oxidation owing to some poor electrical conductivity of the material. When the same material was coated with an overlayer of WO_3 nanosheets, the PEC current increased dramatically. The photocurrent increased even further on depositing $Co_{0.8}Mn_{0.2}O_x$ nanoparticles on the films, which served as co-catalyst. Such samples yielded photocurrent as high as 6.02 mA/cm^2 at 1.23 V (vs. RHE). Additionally, with WO_3 coating onto BBNO, the photoanodes were found to be more stable. In another study [64] Kim et al. observed significant gain in PEC current by using a ternary hybrid material system, represented by CdS-TiO_2-WO_3, as electrode. Enhanced activity in this case was due to swift transfer of photogenerated electrons to WO_3 from CdS via TiO_2, which was assisted through the potential gradient existing at the interface. This eased the separation of charge carriers and reduced recombination. Maximum photocurrent observed under visible light at zero applied potential (vs. Ag/AgCl) was 1.6 mA/cm^2, which was nearly five-times greater than the value recorded with bare CdS. Interface layering of WO_3 with antimony sulphide was studied by Zhang et al. [65]. They first prepared, on FTO glass substrate, WO_3 in varied morphological features by hydrothermal method. Following this, WO_3/SbS_3 heterojunction was evolved that yielded much improvement in PEC current generation for water splitting. Against the photocurrent of 0.45 mA/cm^2, recorded with pristine WO_3, heterojunction films offered maximum photocurrent of 1.79 mA/cm^2 at 0.8 V applied potential (vs. RHE) under simulated sunlight. Suitable energy band positions in the heterostructure, more effective absorption of light and faster separation and transfer of charges mainly contributed to the observed increase in PEC performance.

4.5   Interface layering with TiO_2

TiO_2 is the first reported material for PEC water electrolysis and hydrogen generation [66] and it still continues to be one of the most attractive material options for above application. It is a low-cost, nontoxic material with band edges perfectly straddling the redox levels for the evolution of H_2 and O_2 through PEC water electrolysis. However, since it is a high band gap material with absorption threshold lying in the ultraviolet region, it cannot absorb the bulk of the visible light from the solar spectrum. In order to overcome this limitation, many researchers have attempted interface layering of TiO_2 with other semiconductors, especially the one having low or moderate band gap. This strategy is anticipated to effectively strengthen the absorption in the visible range and also generate faster separation and transmission of photoevolved charge carriers. Momeni and Ghayeb [67] grew ZnO nanotube arrays on titanium foils by electrochemical anodization. Subsequently, ZnO was grafted on it by chemical bath deposition. Thus, obtained ZnO-TiO_2 nanotube arrays were used in PEC water electrolysis. With enhanced surface area at the interface and direct conduction path for charge carriers, generated through aligned nanotubes, a significant gain in PEC photocurrent was recorded. Heterojunction clearly favoured charge separation and effectively reduced recombination losses. Additionally, heterojunction electrodes also exhibited greater stability in contact with electrolyte solution.

Zhang et al. [68] attempted to interface layer TiO_2 with $SrTiO_3$. They first obtained TiO_2 nanotube arrays on titanium sheets through electrochemical anodization, which were then over-layered by $SrTiO_3$ through controlled substitution of Sr hydrothermally. They observed that with increase in $SrTiO_3$ fraction, the flatband potential was shifted to more negative potential thereby significantly influencing PEC water splitting. Nearly 100% rise in external quantum efficiency was recorded and the heterojunction system appeared to be suitable also for quantum-dye-sensitized solar cells.

Interface layering of Cr-doped TiO_2 with indium tin oxide (ITO) was attempted by Dholam et al. [69], which they achieved through deposition of sequential pattern of the two materials by RF sputtering, to have multi-layered thin film structure, that was tested for PEC water splitting. Photocurrent increased significantly in multilayered structure, in comparison with bilayered films, largely attributable to higher absorption of visible light, series of space charge layers in the form of ITO/TiO_2 interfaces, and the generation of photoelectrons in the close vicinity of space charge region that led to significant fall in recombination losses. Sharma et al. [10] prepared a bilayered thin film electrode by interface layering of Fe-doped TiO_2 with Zn-doped Fe_2O_3. The evolved electrodes were used in PEC water splitting and yielded a significant increment in photocurrent density in comparison with the pristine samples of the two oxides.

Interface layering of TiO_2 with TiN has also been studied. Fakhouri et al. employed RF sputtering to prepare multilayered thin films of TiO_2 and TiN. They proposed these as promising systems for PEC water splitting. For 9 and 18 bilayer films, maximum photocurrent was recorded at TiN/TiO_2 ratio 21 and 28%, respectively. The increase in photocurrent clearly evinced efficient separation and migration of photoevolved electrons/holes at the heterojunction [70]. Xu et al. (2011) prepared Bi_2WO_6/TiO_2 heterojunction films and explored their PEC applications. With porous morphology and visible light absorbing characteristics, the films yielded improved PEC response. Significant gain in visible light photocurrent (twice to pristine Bi_2WO_6 and 13-times to pristine TiO_2) was recorded, which resulted from improved charge separation efficiency, elevated interfacial surface area and porous nature of films [71]. In another study, Lin et al. synthesized two-dimensional $TiSi_2$ nanonets by chemical vapor deposition, followed by over-layering with crystalline TiO_2 through atomic layer deposition. The resulting $TiO_2/TiSi_2$ core/shell nanoheterostructures had two very distinct advantages from the viewpoint of their application in PEC water electrolysis: Firstly, TiO_2/electrolyte interfacial area was maximized and secondly with the presence of highly conductive $TiSi_2$, charge separation and transport were facilitated [72]. The two more dominant and photoactive phases of TiO_2, known for photocatalytic applications, are anatase and rutile phases. Li et al. obtained nanoheterojunction of anatase TiO_2/rutile TiO_2 by rapid thermal treatment and explored its role in charge separation, which is one of the most crucial steps involved and often a limiting factor in PEC water splitting. They observed that, in this case, for unidirectional charge transport and effective curb on recombination, appropriate phase alignment was essential. Junction interface with lesser trap states exhibited superior PEC performance with 3 and 9 times increment in photocurrent, respectively, in comparison with the pristine anatase and rutile TiO_2 [73].

5.   Interface layering with carbon materials

Metal oxides constitute the most promising and widely explored class of semiconductors for application in PEC water splitting to produce hydrogen fuel. Although, till date none of the materials investigated possess all the required characteristics, viz. optical, thermodynamic and kinetic, for efficient generation of hydrogen from solar energy [2]. There are various strategies to boost the PEC cell response of semiconductors, which include doping, surface decoration, forming of heterojunction and controlling morphology of nanostructures. Interface layering with carbon materials, viz. amorphous carbon, carbon nanotubes (CNTs) and graphene is another interesting aspect of research being pursued with lot of enthusiasm. A thin coating of carbon layer onto the photoelectrode is known to improve electron transport and separation of charge carriers, thereby imparting greater stability to the electrode along with improved PEC response. Electrically conducting and optically transparent thin layer of carbon can be coated by the simple process of dip-coating the photoelectrode in a suitable carbon precursor followed by carbonization. Interface layering of photoactive semiconductors with CNTs, which are known to exhibit both metallic and semiconducting behavior, is another important modification towards achieving higher efficiency in PEC water splitting. In several studies, CNTs have been used as good catalytic support to grow metal oxide semiconductors, with samples yielding faster transportation of photogenerated charge carriers. Although, it is worth mentioning here that the PEC response of CNTs alone is not very good as these are susceptible to quick agglomeration that hinders the charge conduction process [74]. Graphene is another promising carbon material for applications in PEC water splitting. It is hydrophilic in nature. With its remarkable electrical conductivity, it can serve as sink for photogenerated electrons and thereby play a favourable role in quick separation and migration of photogenerated charge carriers. It has generally been interlayered with metal oxide semiconductors by first using graphene oxide (GO) as precursor, which is subsequently chemically reduced to reduced graphene oxide (rGO). It has also been reported that when combined with semiconductors like ZnO and TiO_2, electrons can easily flow from conduction bands of metal oxides to GO. It is primarily due to the fact that the conduction band energy of above oxides is more negative [75]. Presented below is a brief account of few recent studies performed by different workers on interface layering/over-layering of metal oxide semiconductors with carbon-based materials and their effects on increasing the efficacy of H_2 production by PEC splitting of water.

Hou et al. [76] explored interface layering of p-Cu_2O with n-TaON and used heterojunction films as photoanode. They also attempted surface passivation by interface layering with an ultrathin carbon sheath. Thus, prepared electrodes were not only more stable but also offered higher performance in PEC water electrolysis. In another work, Zhang et al, [77] also observed efficient water splitting when cuprous oxide nanowire array photoelectrodes were protected by a thin coating of carbon layer. Shi et al, [78] also attempted coating of Cu_2O nanowire photoelectrodes, prepared by electrochemical anodization, by a thin layer of carbon. The photoelectrodes were first dipped in glucose solution with subsequent carbonization. The carbon layer interfaced photoelectrodes presented higher stability under illumination and improved PEC water electrolysis. Wang et al, [79] synthesized particulate photocatalyst sheet, $SrTiO_3$:La,Rh/C/$BiVO_4$:Mo, which they employed as Z-scheme-photocatalyst and achieved efficient splitting of water molecule. In this study, they achieved faster electron transfer by using a conducting film of carbon instead of the more commonly used Au and Rh for this purpose. Thus, they successfully demonstrated the use of carbon-based photocatalyst sheet as a low-cost alternative to the conventional and more expensive materials.

Use of carbon nanotubes (CNTs) as an interfacial layer/over-layer with other semiconductors, especially the metal oxides, has been investigated by many workers in the past. A recent review by Rai et al. [74] has very clearly demonstrated the significance of CNTs in obtaining efficient photoactive materials for use in hydrogen production via PEC water electrolysis. Chaudhary et al. [80] attempted to integrate multiwalled carbon nanotubes (MWCNTs) with ZnO for enhanced PEC water splitting. In this study, a very simple chemical method was followed to grow on the surface of MWCNTs the nanoparticles of ZnO. Observed increment in PEC photocurrent on using the prepared film samples as photoanode was ascribed to a more effective charge collection and transport, mediated through diminished band-bending and more effective interfacial electron transfer. In another study, Zhang et al. [81] also investigated ZnO nanowires that were integrated with CNTs for their role in the flow of photoevolved charge carriers. They reported many apparent advantages with the studied material systems that were crucial for their use in PEC water splitting. [82] Kim et al. synthesized MWCNT-CdS hybrids, which they reported as efficient material for photocatalytic hydrogen generation from water. Integration of TiO_2 with MWCNTs has also been attempted by many researchers [83-85]. In all such studies the photocatalytic response of the material for water splitting was reported to be enhanced significantly. Rai et al. [86] attempted interface layering of α-Fe_2O_3 with MWCNTs and observed that it resulted in significant gain in PEC photocurrent density for water splitting. Primary reason for observed gain in current under illumination was ascribed to the role of MWCNTs in increasing charge conduction and reducing recombination.

Lu et al. [87] prepared three-dimensional heterostructures representing TiO_2 modified by MoS_2 and graphene oxide (GO) by hydrothermal method followed by dip coating. Prepared samples yielded enhanced PEC response, which was largely ascribable to superior absorption of visible light and faster charge transmission. TiO_2, which is a wide band gap material mostly absorbing in the UV region, exhibits significant absorption of visible light also when it is inter-layered with a narrow band gap semiconductor MoS_2. However, geometrical configuration of TiO_2/MoS_2 heterostructure is crucial for PEC performance and needs to be optimized for hydrogen generation. With three different systems investigated, the maximum hydrogen production (μmol) recorded at 3h irradiation time was: TiO_2 – 1.72, TiO_2/GO – 15.6, TiO_2/MoS_2 – 33.5. Wang et al. [88] attempted interface layering of rGO modified TiO_2 with Bi_2S_3 and obtained unique nanoheterostructures that showed superior absorption of visible light. Furthermore, on creation of stepwise band-edge structure in this case, an efficient charge separation was also recorded that helped ultimately in PEC water splitting process. Lin et al. [89] attempted interface layering of rGO with WO_3 to obtain WO_3-rGO nanoheterostructure. It enhanced PEC photocurrent for water splitting significantly at applied potential greater than 0.7 V (vs. Ag/AgCl). At lower bias, however, decrease in PEC response, in comparison with bare WO_3, was observed. They concluded that due to occurrence of interface states, rGO has enhanced separation of electrons/holes. But these interface states can also work as recombination centres and may induce the loss of charge carriers. Hence, a large positive bias was required to induce quick separation of photoevolved electrons/holes in this case. In another work by Ng et al. [75] when rGO was combined with $BiVO_4$ to obtain $BiVO_4$-rGO nanocomposite, the resulting samples yielded a marked improvement in photocatalytic ability for water splitting. PEC photocurrent with such samples recorded as high as 10-fold gain against pristine $BiVO_4$. The observed enhancement in photocurrent was majorly ascribable to effective drop in recombination losses of photogenerated charge carriers. Lin et al. [90] explored the fabrication of a buried heterojunction of multilayered graphene micro-net on silicon for use in PEC water splitting. They also employed a thin layer of TiO_2, which was coated over heterojunction through atomic layer deposition to protect the heterojunction from electrolytic corrosion without compromising on efficient movement of holes. TiO_2 layer facilitated the flow of charge carriers through interface. Silicon is a low-cost and earth-abundant promising PEC material that exhibits faster mobility of charge carriers and low band gap. Hence, Si/graphene heterojunction film is an interesting option for PEC water electrolysis. In this study, introduction of a thin film of $FeNiCoO_x$ co-catalyst onto the heterojunction films yielded very high photocurrent (19 mA/cm^2) under one sun illumination and 1.5 V applied bias (vs. RHE). Interface layering of graphene nanoparticles (GNP) with α-Fe_2O_3 was studied by Rai et al. [91]. By exploring the morphological, optical and PEC characteristics of the prepared samples, they observed that GNP on integration with α-Fe_2O_3 was able to sort out substantially the issues of poor charge-conduction and high rate of recombination, which have been typical problems associated with α-Fe_2O_3.

6.   Interface layering with low-cost non-metallic semiconductors

The development of efficient and stable semiconductors for use as photoelectrode in PEC water electrolysis device is still a challenge. In most cases, electron-hole recombination imparts a major constraint on the performance of photocathode. To overcome this limitation, one strategy is to fabricate nanoheterostructures involving more than one material. Along with the development of metal based semiconductor systems for above purpose, in recent years many workers have also worked on non-metallic semiconductors and especially on nanoheterostructures evolved through interface layering of metallic semiconductors with non metallic semiconductors [92-94]. One of the non-metallic semiconductors investigated with lot of interest is g-C_3N_4 (graphitic carbon nitride). It is a quite stable material (both chemically and thermally) having narrow band gap. Further, its band-structure also favours spontaneous splitting of water. However, g-C_3N_4 suffers from high recombination rate [95]. To overcome this limitation, its interfacial layering with metal oxide semiconductors has been attempted by few researchers. Wojtyla et al. [92] attempted to combine copper oxide with g-C$_3$N$_4$ and evolved a Type-II nanoheterostructure that yielded improved PEC performance for water splitting against the neat g-C$_3$N$_4$. The increase in PEC response was mainly attributed to reduced recombination in nanoheterostructure. Further, on using metallic copper underlayer or copper iodide overlayer in the nanoheterostructure, the PEC activity was further boosted. An et al. investigated series of metal oxide nanorod arrays/carbon nitride nanohetrostructures for PEC water oxidation. They used a hydrolyzed and homogeneous suspension of carbon nitride as precursor in the synthesis of following material systems: Fe$_2$O$_3$/C$_3$N$_4$, TiO$_2$/C$_3$N$_4$, ZnO/C$_3$N$_4$ and WO$_3$/C$_3$N$_4$, which were studied as photoanode. Co-Pi co-catalyst was also layered over the surface of heterostructure to enhance further the rate of water oxidation. Significant improvement in PEC activity was recorded, largely attributable to increase in interfacial conductivity and enhanced charge separation [96]. In another study, Hou et al. [97] obtained the heterojunction of WO$_3$ and C$_3$N$_4$ as thin film. Prepared samples were used as photoanode to achieve significant gain in the PEC cell performance towards oxidation of water. The samples were prepared by first growing hydrothermally WO$_3$ nanosheet array on tungsten mesh. Following it, C$_3$N$_4$ coating was applied, over the previously grown WO$_3$ layer, by wet impregnation and sintering. Superior light harvesting, efficient separation-migration of photoevolved electrons/holes and the unique three-dimensional nanoheterostructure of samples contributed mainly to the excellent PEC performance. In an interesting study, Li et al. [98] attempted to prepare, by hydrothermal route followed by electrodeposition, CdS@MoS$_2$ bilayered (core/shell) nanohetrostructured thin films. In this study, CdS nanorods were over-layered by MoS$_2$ and the heterostructures showed improved separation and flow of photoevolved charge carriers, leading to significant gain in PEC photocurrent.

7.   Interface layering/integration with metal nanoparticles

In the overall PEC process for hydrogen production via photoelectrolysis of water, the loss of photogenerated charge carriers via recombination often compete with the reaction rate for hydrogen and oxygen evolution on electrode surfaces. It is a serious limitation with many metal oxide semiconductors that adversely affect their PEC performance. To overcome the above hurdle, interface layering/integration of metal oxide semiconductors with plasmonic nanostructures/nanoparticles of metals has also been attempted by few workers with very promising outcome. With a proper tuning of metal nanoparticles led surface plasmon resonance, the range of wavelength for light absorption by materials can

Photoelectrochemical Water Splitting: Materials and Applications   Materials Research Forum LLC
Materials Research Foundations **70** (2020) 58-90   https://doi.org/10.21741/9781644900734-3

be expanded. This effect is particularly useful with high band gap materials like TiO_2 and ZnO to extend absorption to the visible range. Furthermore, metal nanoparticles also serve as electron acceptors and thus play a vital role in electron-hole pair separation. There are few studies reported in literature that highlight the significance of this strategy in enhancing the efficacy of PEC water electrolysis. Yao et al. [99] surface-decorated gold nanoparticles onto hierarchical bilayered nanoheterostructure of high band gap TiO_2 and low band gap Bi_2WO_6. They observed Plasmonic effect induced significant enhancement in the PEC response under visible light. They fabricated $TiO_2/Bi_2WO_6/Au$ heterojunction nanorod arrays through hydrothermal and solvothermal routes. In another investigation Wei et al. [100] attempted interface layering of ZnO/CdS nanoarrays with gold nanoparticles to obtain ZnO/CdS/Au. Prepared samples were employed as photoanode in PEC water electrolysis. With a favourable band alignment at ZnO/CdS junction and the surface plasmon resonance induced by Au nanoparticles, prepared samples showed improved PEC response, with photocurrent density reaching as high as 21.53 mA/cm^2 under visible light and 1.2 V applied potential (vs. Ag/AgCl). The role of plasmonic layer in enhancing PEC splitting of water, using α-Fe_2O_3 photoelectrode, was investigated by Verma et al. [3]. Owing to poor electrical conduction and high rate of recombination, α-Fe_2O_3, which is otherwise, a material with almost perfect bandgap for PEC splitting of water, suffers adversely in overall PEC photocurrent generation. However, with the overlayering of metal nanoparticles on its surface, electron-hole separation is facilitated, leading to substantial gain in photocurrent.

Conclusion and future directions

The conversion of solar energy to high energy chemical fuels, like hydrogen, via PEC water electrolysis, would continue to be a core area of research in coming years. Material science has to play the most significant and pivotal role in this research. The availability, cost, efficiency of energy conversion and stability are the major parameters on which all possible material-options for this application would be evaluated. Semiconductor nanohetrostructures/nanoarchitectures, evolved through thoughtful integration of different semiconductors by interface layering/over-layering have shown high promise for hydrogen generation through PEC water electrolysis. Such systems, emerging as strong class of materials for above application, deserve more detailed investigations. With the rapid development in advanced synthesis routes and better characterization tools, significant advancements in knowledge and capacity building, to this most vital area of research, are expected in near future.

Acknowledgements

Authors are grateful to University Grants Commission, Government of India for financial assistance under research project (F. No. 194-1/2009 (IC); July 2015).

References

[1] B. Kumari, S. Sharma, N. Singh, A. Verma, V.R. Satsangi, S. Dass, R. Shrivastav, ZnO thin films, surface- embedded with biologically derived Ag/Au nanoparticles, for efficient photoelectrochemical splitting of water, Int. J. Hydrogen Energy 39 (2014) 18216-18229. https://doi.org/10.1016/j.ijhydene.2014.09.025

[2] V. Sharma, M. Dixit, V.R. Satsangi, S. Dass, S. Pal, R. Shrivastav, Photoelectrochemical splitting of water with nanocrystalline $Zn_{1-x}Mn_xO$ thin films: First-principle DFT computations supporting the systematic experimental endeavour, Int. J. Hydrogen Energy 39 (2014) 3637-3648. https://doi.org/10.1016/j.ijhydene.2013.12.150

[3] A. Verma, A. Srivastav, A. Banerjee, D. Sharma, S. Sharma, U.B. Singh, V.R. Satsangi, R. Shrivastav, D.K. Avasthi, S. Dass, Plasmonic layer enhanced photoelectrochemical response of Fe_2O_3 photoanodes, J. Power Sources 315 (2016) 152-160. https://doi.org/10.1016/j.jpowsour.2016.03.004

[4] D. Sharma, S. Upadhyay, V.R. Satsangi, R. Shrivastav, U.V. Waghmare, S. Dass, Ni-doped Cu_2O thin films for solar-hydrogen generation: Experiments & first-principles analysis, Adv. Sci. Lett. 22 (2016) 780-784. https://doi.org/10.1166/asl.2016.6920

[5] S. Choudhary, A. Solanki, D. Sharma, N. Singh, S. Upadhyay, R. Shrivastav, V.R. Satsangi, S. Dass, Photoelectrochemical water splitting using bilayered $ZnO/SrTiO_3$ photoelectrodes, Int. J. Modern Phys. 22 (2013) 545-551. https://doi.org/10.1142/S2010194513010647

[6] S. Choudhary, A. Solanki, S. Upadhyay, N. Singh, V.R. Satsangi, R. Shrivastav, S. Dass, Nanostructured $CuO/SrTiO_3$ bilayered thin films for photoelectrochemical water splitting, J. Solid State Electrochem. 17 (2013) 2531-2538. https://doi.org/10.1007/s10008-013-2139-7

[7] A. Ikram, S. Sahai, S. Rai, S. Dass, R. Shrivastav, V.R. Satsangi, Improved charge transportation at PbS QDs/TiO_2 interface for efficient PEC hydrogen generation, Phys. Chem. Chem. Phys. 18 (2016) 15815-15821. https://doi.org/10.1039/C6CP00854B

[8] A.G. Tamirat, J. Rick, A.A. Dubale, W.N. Sub, B.J. Hwang, Using hematite for photoelectrochemical water splitting: a review of current progress and challenges, Nanoscale Horizon 1 (2016) 243-267. https://doi.org/10.1039/C5NH00098J

[9] P. Kanhere, Z. Chen, A review on visible light active perovskite-based photocatalysts, Molecules 19 (2014) 19995-20022. https://doi.org/10.3390/molecules191219995

[10] P. Sharma, P. Kumar, A. Solanki, R. Shrivastav, S. Dass, V.R. Satsangi, Photoelectrochemical performance of bilayered $Fe-TiO_2/Zn-Fe_2O_3$ thin films for solar generation of hydrogen, J. Solid State Electrochem. 16 (2012) 1305-1312. https://doi.org/10.1007/s10008-011-1552-z

[11] B.Y. Cheng, J.S. Yang, H. W. Cho, J.J. Wu, Fabrication of efficient $BiVO_4/TiO_2$ heterojunction photoanode for photoelectrochemical water oxidation, ACS Appl. Mater. Interfaces 8 (2016) 20032-20039. https://doi.org/10.1021/acsami.6b05489

[12] H. Dong, Q. Liu, Y. He, Preparation of nanoporous $BiVO_4/TiO_2/Ti$ film through electrodeposition for photoelectrochemical water splitting, R. Soc. open sci. 5(9) 180728 (1-10).

[13] J. Resasco, H. Zhang, N. Kornienko, N. Becknell, H. Lee, J. Guo, A. L. Briseno, P. Yang, $TiO_2/BiVO_4$ nanowire heterostructure photoanodes based on Type II band alignment, ACS Cent. Sci. 2 (2016) 80-88. https://doi.org/10.1021/acscentsci.5b00402

[14] J. Su, L. Guo, N. Bao, C. A. Grimes, Nanostructured $WO_3/BiVO_4$ heterojunction films for efficient photoelectrocehmical water splitting, Nano Lett. 11 (2011) 1928–1933. https://doi.org/10.1021/nl2000743

[15] M.G. Mali, H. Yoon, M. W. Kim, M. T. Swihart, S.S. Al-Deyab, S. S. Yoon, Electrosprayed heterojunction $WO_3/BiVO_4$ films with nanotextured pillar structure for enhanced photoelectrochemical water splitting, Appl. Phys. Lett. 106 (2015) 151603 (1-5). https://doi.org/10.1063/1.4918583

[16] L. Yan, W. Zhao, Z. Liu, 1D $ZnO/BiVO_4$ heterojunction photoanodes for efficient photoelectrochemical water splitting, Dalton Trans. 45 (2016) 11346-11352. https://doi.org/10.1039/C6DT02027E

[17] M. Long, W. Cai, H. Kisch, Visible light induced photoelectrochemical properties of n-$BiVO_4$, and n-$BiVO_4$/p-Co_3O_4, J. Phys. Chem. C 112 (2008) 548-554. https://doi.org/10.1021/jp075605x

[18] T. Soltani, A. Tayyebi, B. K. Lee, $BiFeO_3/BiVO_4$ p-n heterojunction for efficient and stable photocatalytic and photoelectrochemical water splitting under visible light irradiation, Catal. Today (2018) 1-32. https://doi.org/10.1021/jp075605x

[19] Y. Chen, M. Yang, J. Du, G. Ke, X. Zhong, Y. Zhou, F. Dong, L. Bian, H. He, $MoO_3/BiVO_4$ heterojunction film with oxygen vacancies for efficient and stable photoelectrochemical water oxidation, J. Mater. Sci. 54 (2019) 671-682. https://doi.org/10.1007/s10853-018-2863-6

[20] U. Shaislamov, K. Krishnamoorthy, S. J. Kim, A. Abidov, B. Allabergenov, S. Kim, S. Choi, R. Suresh, W. M. Ahmed, H. J. Lee, Highly stable hierarchical p-CuO/ZnO nanorod/nanobranch photoelectrode for efficient solar energy conversion, Int. J. Hydrogen Energy 41 (2016) 2253-2262. https://doi.org/10.1016/j.ijhydene.2015.10.146

[21] Q. Zhang, D. Xu, X. Zhou, K. Zhang, Solar hydrogen generation from water splitting using ZnO/CuO hetero nanostructures, Energy Procedia 61 (2014) 345-348. https://doi.org/10.1016/j.egypro.2014.11.1121

[22] J. Kwon, H. Cho, J. Jung, H. Lee, S. Hong, J. Yeo, S. Han, S. H. Ko, ZnO/CuO/M (M = Ag, Au) hierarchical nanostructure by successive photoreduction process for solar hydrogen generation, Nanomaterials 8 (2018) 323 (1-8). https://doi.org/10.3390/nano8050323

[23] A. Kargar, Y. Zing, S. J. Kim, C.T. Riley, X. Pan, D. Wang, ZnO/CuO heterojunction branched nanowires for photoelectrochemical hydrogen generation, ACS Nano 7 (2013) 11112-11120. https://doi.org/10.1021/nn404838n

[24] S.S. Wilson, J. P. Bosco, Y. Tolstova, D.O. Scanion, G.W. Watson, H.A. Atwater, Interface stoichiometry control to improve device voltage and modify band alignment in ZnO/Cu_2O heterojunction solar cells, Energy Environ. Sci. 7 (2014) 3606-3610. https://doi.org/10.1039/C4EE01956C

[25] F. Wu, F. Cao, Q. L, L. Li. Enhancing photoelectrochemical activity with three-dimensional p-CuO/n-ZnO junction photocathodes, Sci. China Mater. 59(10) (2016) 825-832. https://doi.org/10.1007/s40843-016-5054-6

[26] S. Siol, J. C. Hellmann, S.D. Tilley, M. Graetzel, J. Morasch, J. Deuermeier, W. Jaegermann, A. Klein, Band alignment engineering at Cu_2O/ZnO interfaces, ACS Appl. Mater. Interfaces 8 (2016) 21824-21831. https://doi.org/10.1021/acsami.6b07325

[27] M. Izaki, T. Shinagawa, K.T. Mizuno, Y. Ida, M. Inaba, A. Tasaka, Electrochemically constructed $p-Cu_2O/n-ZnO$ heterojunction diode for photovoltaic device, J. Phys. D: Appl. Phys. 40 (2007) 3326-3329. https://doi.org/10.1088/0022-3727/40/11/010

[28] E.L. Tsege, S.K. Cho, L. T. Tufa, V.V. Tran, J. Lee, H.K. Kim, Y.H. Hwang, Scalable and inexpensive strategy to fabricate CuO/ZnO nanowire heterojunction for efficient photoinduced water splitting, J. Mater. Sci. 53 (2017) 2725-2734. https://doi.org/10.1007/s10853-017-1711-4

[29] C. Liu, F. Meng, L. Zhang, D. Zhang, S. Wei, K. Qi, J. Fan, H. Zhang, X. Cui, CuO/ZnO heterojunction nanoarrays for enhanced photoelectrochemical water Oxidation, Appl. Surf. Sci. 469 (2019) 276-282. https://doi.org/10.1016/j.apsusc.2018.11.054

[30] G. Dong, B. Du, L. Liu, W. Zhang, Y. Liang, H. Shi, W. Wang, Synthesis and their enhanced photoelectrochemical performance of ZnO nanoparticle-loaded CuO dandelion heterostructures under solar light, Appl. Surf. Sci. 399 (2016) 86-94. https://doi.org/10.1016/j.apsusc.2016.12.024

[31] S. Zhang, X.B. Cao, J. Wu, L. W. Zhu, L. Gu, Preparation of hierarchical CuO@TiO$_2$ nanowire film and its application in photoelectrochemical water splitting, Trans. Nanoferrous Met. Soc. China 26 (2016) 2094-2101. https://doi.org/10.1016/S1003-6326(16)64324-8

[32] T.S. Atabaev, D.H. Lee, N.H. Hong, Fabrication of TiO$_2$/CuO photoelectrode with enhanced solar water splitting activity, Functional Mater. Lett. 10 (2017) 1750084 (1-4). https://doi.org/10.1142/S1793604717500849

[33] Z. Li, J. Liu, D. Wang, Y. Gao, J. Shen, Cu$_2$O/Cu/TiO$_2$ nanotube Ohmic heterojunction arrays with enhanced photocatalytic hydrogen production activity, Int. J. Hydrogen Energy 37 (2012) 6431-6437. https://doi.org/10.1016/j.ijhydene.2012.01.075

[34] G.K. Mor, O. K. Varghese, R.H.T. Wilke, S. Sharma, K. Shankar, T.J. Latempa, K.S. Choi, C.A. Grimes, p-type Cu-Ti-O nanotube arrays and their use in self-biased heterojunction photoelectrochemical diodes for hydrogen generation, Nano Lett. 8 (2008) 1906-1911. https://doi.org/10.1021/nl080572y

[35] W. Siripala, A. Ivanoskaya, T.F. Jaramillo, S.H. Baeck, E.W. McFarland, A Cu$_2$O/TiO$_2$ heterojunction thin film cathode for photoelectrocatalysis, Solar Ener. Mater. Solar Cells 77 (2003) 229-237. https://doi.org/10.1016/S0927-0248(02)00343-4

[36] M.G. Mendez-Medrano, E. Kowalska, A. Lehoux, A. Herissan, B. Ohtani, D. Bahena, V. Briois, C. Colbeau-Justin, J.L. Rodriguez-Lopez, H. Remita, Surface modification of TiO$_2$ with Ag nanoparticles and CuO nanoclusters fopr application in photocatalysis, J. Phys. Chem. C 120 (2016) 5143-5154. https://doi.org/10.1021/acs.jpcc.5b10703

[37] S.J.A. Moniz, J. Tang, Charge transfer and photocatalytic activity in CuO/TiO_2 nanoparticle heterojunctions synthesized through a rapid, one-pot, microwave solvothermal route, ChemCatChem. 7 (2015) 1659-1667. https://doi.org/10.1002/cctc.201500315

[38] D. Sharma, S. Upadhyay, R. Shrivastav, V.R. Satsangi, U. V. Waghmare, S. Dass, Nanostructured $BaTiO_3/Cu_2O$ heterojunction with improved photoelectrochemical activity for H_2 evolution: Experimental and first-principles analysis, Appl. Catal. B: Environ. 189 (2016) 75-85. https://doi.org/10.1016/j.apcatb.2016.02.037

[39] P. Samarasekara, Characterization of low cost $p-Cu_2O/n-CuO$ junction, GESJ Physics 2 (4) (2010) 3-8.

[40] A. Verma, A. Srivastava, S. Sharma, P. Badami, V.R. Satsangi, R. Srivastava, A.M. Kannan, D.K.C. Awasthi, S. Dass, MWCNTs and Cu_2O sensitized $Ti-Fe_2O_3$ photoanode for improved water splitting, Int. J. Hydrogen Energy 43 (2018) 6049-6059. https://doi.org/10.1016/j.ijhydene.2018.01.204

[41] J. Juodkazyte, B. Sabeka, I. Savickaja, V. Pakstas, A. Naujokaitis, A. Griguceviciene, Study on charge transfer processes in thin-film heterojunction between cuprous oxide and hematite, Mater. Sci. Semiconductor Processing 80 (2018) 56-62. https://doi.org/10.1016/j.mssp.2018.02.020

[42] D. Sharma, S. Upadhyay, A. Verma, V.R. Satsangi, R. Shrivastav, S. Dass, Nanostructured $Ti-Fe_2O_3/Cu_2O$ heterojunction photoelectrode for efficient hydrogen production, Thin Solid Films 574 (2015) 125-131. https://doi.org/10.1016/j.tsf.2014.12.003

[43] D. Sharma, S. Upadhyay, V.R. Satsangi, R. Shrivastav, U.V. Waghmare, S. Dass, Improved photoelectrochemical water splitting performance of $Cu_2O/SrTiO_3$ heterojunction photoelectrode, J. Phys. Chem. C 118 (2014) 25320-25329. https://doi.org/10.1021/jp507039n

[44] D. Sharma, A. Verma, V.R. Satsangi, R. Shrivastav, S. Dass, Nanostructured $SrTiO_3$ thin films sensitized by Cu_2O for photoelectrochemical hydrogen generation, Int. J. Hydrogen Energy 39 (2014) 4189-4197. https://doi.org/10.1016/j.ijhydene.2013.12.201

[45] X. Cheng, J. Ding, Y. Wu, H. Liu, G. Dawson, The photocathodic properties of α-Fe_2O_3 wrapped $CuFeO_2$ layer on ITO glass for water splitting, Chem. Phys. 513 (2018) 241-245. https://doi.org/10.1016/j.chemphys.2018.08.009

[46] R. Rajendran, Z. Yakoob, M. A. M. Teridi, M.S. A. Rahman, K. Sopian, Preparation of nanostructured p-NiO/n-Fe$_2$O$_3$ heterojunction and study of their enhanced photoelectrochemical water splitting performance, Mater. Lett. 133 (2014) 123-126. https://doi.org/10.1016/j.matlet.2014.06.157

[47] A.K. Singh, D. Sarkar, A facile approach for preparing densely-packed p-NiO/n-Fe$_2$O$_3$ heterojunction nanowires for photoelectrochemical water splitting, Nanoscale 10 (2018) 13130-13139. https://doi.org/10.1039/C8NR02508H

[48] Y.W. Phuan, M.N. Chong, J.D. Ocon, E.S. Chan, A novel ternary nanostructured carbonaceous-metal-semiconductor eRGO/NiO/α-Fe$_2$O$_3$ heterojunction photoanode with enhanced charge transfer properties for photoelctrochemical water splitting, Solar Ener. Mater, Solar Cells 169 (2017) 236-244. https://doi.org/10.1016/j.solmat.2017.05.028

[49] Y.W. Phuan, M.N. Chong, T. Zhu, E.S. Chan, J.D. Ocon, Employing electrochemical reduced graphene oxide as a co-catalyst for synergistically improving the photoelctrochemical performance of nanostructured hematite thin films, J. Taiwan Inst. Chem. Engineers 71 (2017) 510-517. https://doi.org/10.1016/j.jtice.2016.12.017

[50] S.K. Mohapatra, S. Banerjee, M. Misra, Synthesis of Fe$_2$O$_3$/TiO$_2$ nanorod-nanotube arrays by filling TiO$_2$ nanotubes with Fe, Nanotechnology 19 (2008) 315601 (1-7). https://doi.org/10.1088/0957-4484/19/31/315601

[51] M. Niu, F. Huang, L. Cui, P. Huang, Y. Yu, Y. Wang, Hydrothermal synthesis, structural characteristics, and enhanced photocatalysis of SnO$_2$/α-Fe$_2$O$_3$ semiconductor nanoheterostructures, ACS Nano 4 (2010) 681-688. https://doi.org/10.1021/nn901119a

[52] C. Chen, H. Bai, Z. Du, M. Li, X. Yan, J. Jiang, W. Fan, W. Shi, Hydrothermal synthesis of Fe$_2$O$_3$/ZnO heterojunction photoanode for photoelctrochemical water splitting, Functional Mater. Lett. 8 (2015) 1550058 (1-5). https://doi.org/10.1142/S1793604715500587

[53] F. Bouhjar, B. Mari, B. Bessais, Hydrothermal fabrication and characterization of ZnO/Fe$_2$O$_3$ heterojunction devices for hydrogen production, J. Anal. Pharm. Res. 7 (2018) 315-321. https://doi.org/10.15406/japlr.2018.07.00246

[54] C. Miao, S. Ji, G. Xu, G. Liu, L. Zhang, C. Ye, Micro-nano-structured Fe$_2$O$_3$:Ti/ZnFe$_2$O$_4$ heterojunction films for water oxidation, ACS Appl. Mater. Interfaces 4 (2012) 4428-4433. https://doi.org/10.1021/am3011466

[55] M.G. Ahmed, T.A. Kandiel, A.Y. Ahmed, I. Kretschmer, F. Rashwan, D. Bahnemann, Enhanced photoelctrochemical water oxidation on nanostructured

hematite photoanodes via p-$CaFe_2O_4$/n-Fe_2O_3 heterojunction formation, J. Phys. Chem. C 119 (2015) 5864-5871. https://doi.org/10.1021/jp512804p

[56] Y. Kawai, K. Negai, T. Abe, A visible-light-induced photoelectrochemical water-splitting system featuring an organo-phtocathode along with a tungsten oxide photoanode, RSC Adv. 7 (2017) 34694-34698. https://doi.org/10.1039/C7RA05272C

[57] A. Muller, I. Kondofersky, A. Folger, D. Fattakhova-Rohlfing, T. Bein, C. Scheu, Dual absorber Fe_2O_3/WO_3 host-guest architectures for improved charge generation and transfer in photoelctrochemical applications, Mater. Res. Express 4 (2017) 016409 (1-9). https://doi.org/10.1088/2053-1591/aa570f

[58] P. Wu, Z. Liu, D. Chen, M. Zhou, J. Wei, Flake-like NiO/WO_3 p-n heterojunction photocathode for photoelectrochemical water splitting, Appl. Surf. Sci. 440 (2018) 1101-1106. https://doi.org/10.1016/j.apsusc.2018.01.292

[59] K.C. Leonard, K.M. Nam, H.C. Lee, S.H. Kang, H.S. Park, A. J. Bard, $ZnWO_4$/WO_3 composite for improving photoelectrochemical water oxidation, J. Phys. Chem. C 117 (2013) 15901-15910. https://doi.org/10.1021/jp403506q

[60] S.F. Anis, B.. Lalia, G. Palmisano, R. Hashaikeh, Photoelectrochemical activity of electrospun WO_3/$NiWO_4$ nanofibers under visible light irradiation, J. Mater. Sci. 53 (2018) 2208-2220. https://doi.org/10.1007/s10853-017-1633-1

[61] H. Zhang, W. Tian, Y. Li, H. Sun, M.O. Tade, S. Wang, Heterostructured WO_3@$CoWO_4$ bilayer nanosheets for enhanced visible-light photo, electro and photoelectron-chemical oxidation of water, J. Mater. Chem. A 6 (2018) 6111-6121. https://doi.org/10.1039/C8TA90082E

[62] A. Martinez-Garcia, V.K. Vendra, S. Sunkara, P. Haldankar, J. Jasinski, M.K. Sunkara, Tungsten oxide-coated copper oxide nanowire arrays for enhanced activity and durability with photoelectrochemical water splitting, J. Mater. Chem. A 1 (2013) 15235-15241. https://doi.org/10.1039/c3ta13912c

[63] B. Weng, C.R. Grice, J. Ge, T. Poudel, X. Deng, Y. Yan, Barium bismuth niobate double perovskite/tungsten oxide nanosheet photoanode for high performance photoelectrochemical water splitting, Adv. Energy Mater. 8(10) (2017) 1-6. https://doi.org/10.1002/aenm.201701655

[64] H. Kim, J. Kim, W. Kim, W. Choi, Enhanced photocatalytic and photelectrochemical activity in the ternary hybrid of CdS/TiO_2/WO_3 through the cascadal electron transfer, J. Phys. Chem. C 115 (2011) 9797-9805. https://doi.org/10.1021/jp1122823

[65] J. Zhang, Z. Liu, Z. Liu, Novel WO_3/Sb_2S_3 heterojunction photocatalyst based on WO_3 of different morphologies for enhanced efficiency in photoelectrochemical water splitting, ACS Appl. Mater. Interfaces 8 (2016) 9684-9691. https://doi.org/10.1021/acsami.6b00429

[66] A. Fujishima, K. Honda, Electrochemical photolysis of water at a semiconductor electrode, Nature 238 (1972) 37-38. https://doi.org/10.1038/238037a0

[67] M.M. Momeni, Y. Ghayeb, Visible light-driven photoelectrochemical water splitting on $ZnO-TiO_2$ heterogenous nanotube photoanodes, J. Appl. Electrochem. 45 (2015) 557-566. https://doi.org/10.1007/s10800-015-0836-x

[68] J. Jhang, J. H. Bang, C. Tang, P.V. Kamat, Tailored $TiO_2-SrTiO_3$ heterostructure nanotube arrays for improved photoelectrochemical performance, ACS Nano 4 (2010) 387-395. https://doi.org/10.1021/nn901087c

[69] R. Dholam, N. Patel, A. Santini, A. Miotello, Efficient indium tin oxide/Cr-doped TiO_2 multilayer thin films for H_2 production by photocatalytic water-splitting, Int. J. Hydrogen Energy 35 (2010) 9581-9590. https://doi.org/10.1016/j.ijhydene.2010.06.097

[70] H. Fakhouri, F.A. Khonsari, A.K. Jaiswal, J. Pulpytel, Enhanced visible light photoactivity and charge separation in TiO_2/TiN bilayer thin films, Appl. Catal. A 492 (2015) 83-92. https://doi.org/10.1016/j.apcata.2014.12.030

[71] Q.C. Xu, D.V. Wellia, Y.H. Ng, R. Amal, T.T.Y. Tan, Synthesis of porous and visible-light absorbing Bi_2WO_6/TiO_2 heterojunction films with improved photoelectrochemical and photocatalytic performances, J. Phys. Chem. C 115 (2011) 7419-7428. https://doi.org/10.1021/jp1090137

[72] Y. Lin, S. Zhou, X. Liu, S. SHeehan, D. Wang, $TiO_2/TiSi_2$ heterostructures for high-efficiency photoelectrochemical H_2O splitting, J. Am. Chem. Soc. 131 (2009) 2772-2773. https://doi.org/10.1021/ja808426h

[73] A. Li, Z. Wang, H. Yin, S. Wang, P. Yan, B. Huang, X. Wang, R. Li, X. Zong, H. Han, C. Li, Understanding the anatase-rutile phase junction in charge separation and transfer in a TiO_2 electrode for photoelectrochemical water splitting, Chem. Sci. 7 (2016) 6076-6082. https://doi.org/10.1039/C6SC01611A

[74] S. Rai, A. Ikram, S. Sahai, S. Dass, R. Shrivastav, V. R. Satsangi, CNT based photoelectrodes for PEC generation of hydrogen: A review, Int. J. Hydrogen Energy 42 (2016) 3994-4006. https://doi.org/10.1016/j.ijhydene.2016.10.024

[75] Y.H. Ng, A. Iwase, A. Kudo, R. Amal, Reducing graphene oxide on a visible-light BiVO$_4$ photocatalyst for an enhanced photoelectrochemical water splitting, J. Phys. Chem. Lett. 1 (2010) 2607–2612. https://doi.org/10.1021/jz100978u

[76] J. Hou, C. Yang, H. Cheng, S. Jiao, O. Takeda, H. Zhu, High-performance p-Cu$_2$O/n-TaON heterojunction nanorod photoanodes passivated with an ultrathin carbon sheath for photoelctrochemical water splitting, Energy Environ. Sci. 7 (2014) 3758-3768. https://doi.org/10.1039/C4EE02403F

[77] Z. Zhang, R. Dua, L. Zhang, H. Zhu, H. Zhang, P. Wang, Carbon-layer-protected cuprous oxide nanowire arrays for efficient water reduction, ACS Nano 7 (2013) 1709-1717. https://doi.org/10.1021/nn3057092

[78] W. Shi, X. Zhang, S. Li, B. Zhang, M. Wang, Y. Shen, Carbon coated Cu$_2$O nanowires for photoelectrochemical water splitting with enhanced activity, Appl. Surf. Sci. 358 (2015) 404-411. https://doi.org/10.1016/j.apsusc.2015.08.223

[79] Q. Wang, T. Hisatomi, Y. Suzuki, Z. Pan, J. Seo, M. Katayama, T. Minegishi, H. Nishiyama, T. Takata, K. Seki, A. Kudo, T. Yamada, K. Domen, Particulate photocatalyst sheets based on carbon conductor layer for efficient Z-scheme pure-water splitting at ambient pressure, J. Am. Chem. Soc. 139 (2017) 1675-1683. https://doi.org/10.1021/jacs.6b12164

[80] D. Chaudhary, S. Singh, V.D. Vankar, N. Khare, ZnO nanoparticles decorated multi-walled carbon nanotubes for enhanced photocatalytic and photoelectrocemical water splitting, Photochem. Photobiol. 351 (2017) 154-161. https://doi.org/10.1016/j.jphotochem.2017.10.018

[81] W.D. Zhang, L.C. Jiang, J.S. Ye, Photoelectrochemical study on charge transfer properties of ZnO nanowires promoted by carbon nanotubes, J. Phys. Chem. C 113 (2009) 16247-16253. https://doi.org/10.1021/jp905500n

[82] Y.K. Kim, H. Park, Light-harvesting multi-walled carbon nanotubes and CdS hybrids: Application to photocatalytic hydrogen production from water, Energy Environ. Sci. 4 (2011) 685-694. https://doi.org/10.1039/C0EE00330A

[83] D. Chaudhary, N. Khare, V.D. Vankar, Ag nanoparticles loaded TiO$_2$/MWCNT ternary nanocomposite: A visible-light-driven photocatalyst with enhanced photocatalytic performance and stability, Ceram. Int. 42 (2016) 15861-15867. https://doi.org/10.1016/j.ceramint.2016.07.056

[84] K. Ouyang, S. Xie, X.O. Ma, Photocatalytic activity of TiO_2 supported on multi-walled carbon nanotubes under simulated solar irradiation, Ceram. Int. 39 (2013) 7531-7536. https://doi.org/10.1016/j.ceramint.2013.03.004

[85] D. Chaudhary, S. Singh, V.D. Vankar, N. Khare, $Ag/TiO_2/CNT$ photoanode for efficient photolelectrochemical water splitting under visible light irradiation, Int. J. Hydrogen Energy 42 (2017) 7826-7835. https://doi.org/10.1016/j.ijhydene.2016.12.036

[86] S. Rai, A. Ikram, S. Sahai, S. Dass, R. Shrivastav, V.R. Satsangi, Photoactivity of MWCNTs modified α-Fe_2O_3 photoelectrode towards efficient solar water splitting, Renewable Energy 83 (2015) 447-454. https://doi.org/10.1016/j.renene.2015.04.053

[87] H. Li, W. Dong, J. Xi, G. Du, Z. Ji, 3D flowerlike TiO_2/GO and TiO_2/MoS_2 heterostructures with enhanced photoelectrochemical water splitting, J. Mater. Sci. 53 (2018) 7609–7620. https://doi.org/10.1007/s10853-018-2051-8

[88] X. Wang, J. Xie, C.M. Li, Architecting smart "umbrella" Bi_2S_3/rGO-modified TiO_2 nanorod array structures at the nanoscale for efficient photoelectrocatalysis under visible light, J. Mater. Chem. A 3 (2015) 1235–1242. https://doi.org/10.1039/C4TA05846A

[89] J. Lin, P. Hu, Y. Zhang, M. Fan, Z. He, C.K. Ngaw, J.S.C. Loo, D. Liao, T.T.Y. Tan, Understanding the photoelectrochemical properties of a reduced oxide WO_3 heterojunction photoanode for efficient solar-light-driven overall water splitting, RSC Adv. 3 (2013) 9330–9336. https://doi.org/10.1039/c3ra40550h

[90] C. Lin, Y. Xiao, L. Zhang, Y. Li, J. J. Delaunay, H. Zhu, Efficient photoelectrochemical water oxidation enabled by an amorphous metal oxide-catalyzed graphene/silicon heterojunction photoanode, Sust. Ener. Fuels. 2 (2018) 663-672. https://doi.org/10.1039/C7SE00504K

[91] S. Rai, A. Ikram, S. Sahai, S. Dass, R. Shrivastav, V.R. Satsangi, Morphological, optical and photoelectrochemical properties of Fe_2O_3-GNP composite thin films, RSC Adv. 4 (2014) 17671-17679. https://doi.org/10.1039/c3ra46840b

[92] S. Wojtyla, K. Szmit, T. Baran, Type II heterostructures: The way towards improved photoelectrochemical activity of graphitic carbon nitride, J. Inorg. Organomet. Polym. 28 (2018) 492–499. https://doi.org/10.1007/s10904-017-0733-3

[93] J.S. Kim, J.W. Oh, S.I. Woo, Improvement of the photocatalytic hydrogen production rate of g-C_3N_4 following the elimination of defects on the surface, Catal. Today 293 (2017) 8-14. https://doi.org/10.1016/j.cattod.2016.11.018

[94] M. Shao, Y. Shao, J. Chai, Y. Qu, M. Yang, Z. Wang, M. Yang, W.F. Ip, C.T. Kwok, X. Shi, Z. Lu, S. Wang, X. Wang, H. Pan, Synergistic effect of 2D Ti_2C and g-C_3N_4 for efficient photocatalytic hydrogen production, J. Mater. Chem. A 5 (2017) 16748-16756. https://doi.org/10.1039/C7TA04122E

[95] G. Dong, Y. Zhang, Q. Pan, J. Qiu, J. A fantastic graphitic carbon nitride (g-C_3N_4) material: Electronic structure, photocatalytic and photoelectronic properties, Photochem. Photobiol. C 20 (2014) 33-50. https://doi.org/10.1016/j.jphotochemrev.2014.04.002

[96] A. Xiaoqiang, H. Chengzhi, L. Huachun, L. Huijuan, Q. Jiuhui, Strongly coupled metal oxide/reassembled carbon nitride/Co-Pi heterostructures for efficient photoelectrochemical water splitting, ACS Appl. Mater. Interfaces 10 (2018) 6424–6432. https://doi.org/10.1021/acsami.8b01070

[97] H. Yang, Z. Fan, P.D. Alexander, L. Jikai, F. Pingyun, Branched WO_3 nanosheet array with layered C_3N_4 heterojunctions and CoO_x nanoparticles as a flexible photoanode for efficient photoelectrochemical water oxidation, Adv. Mater. 26 (2014) 5043–5049. https://doi.org/10.1002/adma.201401032

[98] Q. Li, F. Shan, B. Sun, Y. Song, F. Wang, J. Ji, Photo-assisted electrocatalysis of CdS-MoS_2 hybrid for hydrogen evolution reaction: Morphology-dependent photoelectroactivity of p-n junction photocathode under bias potential, Int. J. Hydrogen Energy 42 (2017) 5549-5559. https://doi.org/10.1016/j.ijhydene.2016.05.081

[99] L. Yao, W. Wang, Y. Liang, J. Fu, H. Shi, Plasmon-enhanced visible light photoelectrochemical and photocatalytic activity of gold nanoparticle-decorated hierarchical TiO_2/Bi_2WO_6 nanorod arrays, Appl. Surf. Sci. 469 (2019) 829-840. https://doi.org/10.1016/j.apsusc.2018.11.031

[100] R.B. Wei, P.Y. Kuang, H. Cheng, Y.B. Chen, J.Y. Long, M.Y. Zhang, Z.Q. Liu, Plasmon-enhanced photoelectrochemical water splitting on gold nanoparticle decorated ZnO/CdS nanotube arrays, ACS Sustainable Chem. Eng. 5 (2017) 4249-4257. https://doi.org/10.1021/acssuschemeng.7b00242

Photoelectrochemical Water Splitting: Materials and Applications   Materials Research Forum LLC
Materials Research Foundations **70** (2020) 91-109   https://doi.org/10.21741/9781644900734-4

Chapter 4

Narrow Bandgap Semiconductors for Photoelectrochemical Water Splitting

Sonal Singh[1*], Rishabh Sharma[2*] and Manika Khanuja[3]

[1]Shaheed Rajguru College of Applied Sciences for Women, University of Delhi, New Delhi 110096, India

[2]Thin Film Laboratory, Department of Physics, Indian Institute of Technology, New Delhi-110016, India.

[3]Centre for Nanoscience and Nanotechnology, Jamia Millia Islamia (A Central University), New Delhi 110025, India

[1]sonal.singh0811@gmail.com,[2]rishabh.rammstien@gmail.com, [3]manikakhanuja@gmail.com

*Both the authors have contributed equal to this work

Abstract

With the aim of directing research towards the area of water splitting using photoelectrochemical (PEC) cell, it is necessary to optimize the semiconductor materials used as photoelectrodes in PEC system. Narrow band gap materials prove to ace the race due to their several favorable properties towards solar absorption and water splitting. Also, their energy and potential dynamics seem suitable for this particular application. Although, many narrow band gap materials are available to cause complete splitting of water, certain shortcomings limit their full potential efficiency. To overcome these, certain alterations or modifications are required through some strategies in the form of doping, composites, heterojunctions, and coupling etc. Coupling of narrow band gap materials with wide bandgap materials has proved extremely beneficial for both the counterparts in overcoming the drawbacks of both the materials in a composite and ultimately improving the overall efficiency of the PEC device.

Keywords

Narrow Band Gap Materials, Photoelectrochemical, Water Splitting, Coupling

Contents

1.   Introduction

In today's era of depleting fuel resources and exhausting reserves, associated challenges have taken a dramatic turn that has adversely affected the environment. This has caused society to widely accept the need of sustainable fuel primarily based on natural resources; essentially solar energy generated through various effective and productive techniques [1]. Photoelectrochemical splitting of water offers the possibility to utilize sunlight to chemical fuel production through artificial photosynthesis in the form of hydrogen. It is a cost-effective and energy-saving process of attaining solar-to-hydrogen pathway. The use of semiconductor materials to achieve solar water splitting holds great significance. Choices of semiconductor materials and controlling their properties have been the primary concerns to develop highly stable and efficient water splitting PEC devices. Various methods have been adopted to overcome these challenges to an extent. Crystallinity, dimensionality, pressure and temperature, band gap, size, pH dependency and light are some of the factors of nanomaterials that affect the performance of the PEC cell. Commercially used semiconductor materials such as TiO_2 and ZnO are not sufficient alone to cause complete and efficient splitting of water. They are wide band gap materials requiring sunlight in the UV range of solar spectrum. They generally suffer from the problems of narrow photo-absorption range, instability and high recombination rate of e-h pairs. Therefore, some sort of modifications is required in order to create an effective PEC device by addressing these obstacles. Coupling materials with narrow bandgap materials has been proved to be an effective strategy to improve the photoresponse of the material.

This chapter commences with brief introduction about the fundamental aspect of narrow band gap materials as a strategy to improve photoresponse of the material using PEC water splitting process. This further places emphasis on various aspects and factors of the materials that affect the efficiency to optimize the design concept of the PEC cell. First, narrow bandgap materials as a potential candidate for water splitting, alone or in combination with other materials, will be described. Secondly, some promising narrow band gap materials are intended to be described along with their characteristics, technical remarks and advantages as well as disadvantages as a PEC water-splitting materials.

2.   Narrow band gap materials as a strategy to improve photoresponse of the material

Narrow band gap materials are efficient enough to cause splitting of water due to their suitable energy dynamics. Many researchers have shown materials like CuO, Fe_2O_3, BiOI, Bi_2S_3 etc. to exhibit PEC and photocatalytic activity in their pristine forms. However, their yield is still not sufficient due to certain drawbacks. Narrow band gap materials have been widely used to couple with wide band gap materials to overcome their own known shortcomings. Wide band gap possesses insufficient energy to create e-h pair to cause electrons to jump from valence band to conduction band. Such heterostructure system between two semiconductor aids in controlling the interparticle charge transfer by offering favorable relative energy levels. Such coupling of narrow and large band gap semiconductor materials provides a convenient and suitable means to extend the photoresponse of the material by improving its charge transfer properties. Narrow band gap materials offer advantage over large band gap materials in terms of high photoadsorption activity which increases their catalyst activity. The smaller bandgap shows good results in releasing of reactive oxygen species. Recently, narrow bandgap materials have garnered greatly expanded interest as photo-materials. Growing interests in nanostructures and modern epitaxial techniques have bestowed area of applications for some of the unique properties possessed by narrow bandgap materials which make them the materials of choice.

2.1   Bismuth sulfide (Bi_2S_3)

Bismuth sulfide has a narrow band gap of ~1.7 eV which makes it one of the most sought after choices as good candidate in category of narrow band gap materials for various applications. For several years, it has been used extensively as hydrogen sensors, for electrochemical hydrogen storage, biomolecule detection and X-ray computed tomography imaging [2]. More recently, it has now found application in the fields of photoelectrochemical and photocatalysis devices as photoresponsive materials. Fig 1(a) shows the unit cell of bulk Bi_2S_3 crystal structure, at zero pressure and Fig 1 (b) presents first principle band structure for Bi_2S_3. It is considered ideal as light-harvesting substrate, since its absorption reasonably overlaps with the visible and near-infrared part of the solar spectrum with absorption coefficient of the order 10^4 to 10^5 cm^{-1}, and a reasonable IPCE (incident photon to electron conversion efficiency) of ~5% [3,4]. Tahir et al. were the first to demonstrate the synthesis of Bi_2S_3 by single step deposition using chemical vapor deposition technique. Other methods such as solvothermal, solventless and surfactant-assisted approaches, microwave irradiation etc. require complex manipulations and instrumentations and suffer from high costs of production [4]. Reports have shown to couple Bi_2S_3 with many other semiconductor materials in order to form various kinds of configurations including composites, heterojunctions systems and core-shell structures such as Bi_2O_3/Bi_2S_3, $BiVO_4/Bi_2S_3$, Bi_2S_3/WO_3, Bi_2S_3/TiO_2, $CdS-Bi_2S_3$, Bi_2S_3/Bi_2WO_6 to exhibit improved PEC activity [2,5–9]. Bi_2S_3/Bi_2WO_6 heterojunction synthesis reported by Zhang has proved beneficial for separation and transportation of charge carriers. Bi_2S_3 nanocrystallites were coupled with nanostructured SnO_2 films to perform photoelectrochemical studies. To enhance the efficiency outcome of Bi_2S_3-based photoelectrochemical cells, it is essential to optimize the electrolyte, redox couple, and film thickness [3]. Remarkable photoactivity has also been exhibited by core-shell structures [10–12].

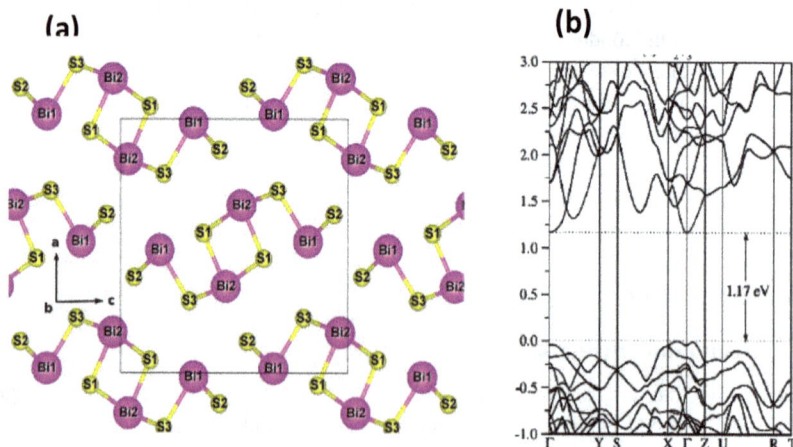

Figure 1: (a) Unit cell of bulk Bi_2S_3 crystal structure, at zero pressure and (b) First principle band structures for Bi_2S_3 [57].

Other than semiconductor materials, use of organic compounds such as carbon nitride and graphene have also been well reported to enhance the photoresponse of the material. Rong et al. reported the use of Bi_2S_3 as modifier coupled with g-C_3N_4 to overcome the low photocatalytic efficiency of g-C_3N_4 alone [13]. Research by Okoth et al. has shown improved and promoted photocurrent response of Bi_2S_3 nanorods by graphene doping [14]. For comparison, methods of deposition, photocurrent performances, and measurement conditions reported for various doped/undoped/composite Bi_2S_3 electrodes have been reviewed in Table 1.

Table 1. Summary of the Photocurrent Performance for Undoped/doped/composite Bi₂S₃ Electrodes Reported in Literature

Material	Authors/year	Deposition method	Photocurrent density	Electrolyte	Illumination source
Bi_2S_3	Mane et al. [60]	Chemical bath deposition	1000 μA /cm^2	0.25 M polysulphide	Tungsten filament lamp
Bi_2S_3 Nanotube	Tahir et al. [4]	AACVD	1.9 mA/cm^2	3 M KCl	Halogen lamp
Bi_2S_3/TiO_2	Brahimi et al. [61]	Hydrothermal process	3×10^{-4} mA/cm^2	1 M KOH	200 W tungsten lamp
Bi_2S_3/TiO_2	Zeng et al. [62]	Chemical route	2 mA/cm^2	0.25 M Na_2S + 0.125 M Na_2SO_3	Xe lamp
CdS-Bi_2S_3	Gadave et al. [63]	Precipitation method	Isc= - 0.2 mA/ cm^2	1 M (NaOH-S-Na_2S)	500 Watt filament lamp

2.2 CuO

CuO is one of the most sought semiconductor metal oxides used for photocathodic hydrogen production. CuO is composed of chained units of CuO_2 and Cu is bonded with oxygen via four coplanar bonds. CuO has a square planar coordination of copper by oxygen in the monoclinic structure and thus stands unique amongst the mono oxides of 3d transition series elements. It is mainly made up from Cu^{2+}-3d wave functions with a = 4.684 Å, b = 3.425 Å, c = 5.129 Å and β = 99.28° as its lattice parameters. Its conduction band and valence band are reported to lie at 4.07 and 5.42 eV below the vacuum level, respectively [15]. Table 2 lists the physical properties of CuO [16,17]. Certain factors such as relatively low conductivity, poor surface evolution kinetics and short diffusion length limit the potential use of C/CuO for high practical solar-hydrogen generation efficiency than their predicted theoretical efficiency values [18–20]. Therefore, in order to improve the photoefficiency, certain methods have to be adopted to modify their

Photoelectrochemical Water Splitting: Materials and Applications   Materials Research Forum LLC
Materials Research Foundations **70** (2020) 91-109   https://doi.org/10.21741/9781644900734-4

optical and electronic properties to increase its conductivity. Strategies involving formation of heterojunctions, composites and use of dopants have been well studied to decrease the resistance to charge carrier transport. Kargar et al. have showed the formation of 3D ZnO/CuO nanowires photocathodes for conversion of solar to hydrogen. They exhibited improved photocathodic current for samples with denser and longer CuO nanowires as nanostructured metal oxide formation helps in improving the surface evolution kinetics by reducing the carrier diffusion length and increasing the surface area of the material [21]. Also, vertical metal oxide nanowires have been beneficial as they provide improved charge collection efficiency, reduced carrier recombination, and enhanced light absorption [22,23]. Furthermore, copper oxides when coupled with stable materials like TiO_2 in heterostructure can enhance their stability by protecting the electrodes surface against decomposition/corrosion in the electrolyte [24]. Guo et al. demonstrated the use of noble metal (Pd) with CuO in improving the semiconductor's HER (Hydrogen Evolution Reaction) efficiency by reducing the overpotential for HER. Previous reports have shown the use of noble metals such as Pt, Rh and Pd as cocatalyst to form "semiconductor oxide/noble metal" systems to improve PEC efficiency [25]. Since CuO also suffers from instability under illumination, HER photocurrent decays quickly with time at CuO photocathodes. Depositing noble metals on the surface of CuO is believed to be a good solution to enhance the stability of CuO [26,27]. Chiang et al. demonstrated the doping of CuO by lithium to investigate its photoelectrochemical properties. It was found that after introducing Li into CuO films, photocurrent density and photon to hydrogen conversion efficiency inflated by five to ten times. Electrical conductivity of the film increased by almost two orders of magnitude as Li^{2+} ion has similar radius to Cu^{2+} ion [28]. Various other examples of CuO heterojunctions for PEC application include Cu_2O/CuO, $CuO/CuWO_4$, $CuO/CuBi_2O_4$, $CuO/SrTiO_3$ [29–32]. Table 3 compares various doped/undoped/composite CuO for their photoelectrochemical properties.

Table 2. Key features and structural properties of CuO

Properties	Value
Density	6.31 g/cm^3
Crystal structure	Monoclinic
Molar mass	79.545 g/mol
Melting point	1200 °C
Boiling point	2000°C
Dielectric constant	18.1
Bang gap (E_g)	Direct 1.21-1.55 eV
Cell content	4 [CuO]
Cell volume	81.08 Å
Solubility in water	Insoluble

Table 3. Summary of the Photocurrent Performance for Undoped/doped/composite CuO Electrodes Reported in Literature

Material	Authors/ year	Deposition method	Photo-current density	Potential	Electrolyte	Illumination source
CuO	Chauhan et al. [15]	Sol-gel	2.3 mA/cm^2	-	0.1 M NaOH	150 W Xenon Arc lamp
ZnO/CuO	Kargar et al. [21]	Thermal oxidation/ hydrothermal growth	-	-0.5 V versus Ag/AgCl RE	0.25M Na$_2$SO$_4$	150 W xenon lamp, 1.5 AM
CuO/Pd	Guo et al. [25]	Thermal oxidation process	1.40 mA/cm^2	0.1 V vs SCE	0.2 M H$_2$BO$_3^-$ /H$_3$BO$_3$	500 W Xenon lamp
Cu$_2$O/CuO/Ni	Dubale et al. [32]	Electrolysis deposition	-2.1 mA/cm^2	0 V vs. RHE	1M Na$_2$SO$_4$	300 W Xenon lamp, 1.5 AM
Li doped CuO	Chiang et al. [28]	Spray pyrolysis	1.69 mA/cm^2	-0.55 V vs. Ag/AgCl	1M KOH	150 W solar simulator

2.3   Fe$_2$O$_3$

Hematite (α-Fe$_2$O$_3$) is the cheap and abundantly available n-type narrowband gap (2-2.2 eV) material for water splitting application as it sufficiently utilizes about 40% of the sunlight incident upon material. Fig 2 shows the crystal structure of α-Fe$_2$O$_3$. It is highly stable in most electrolytes at pH>3 and is environmentally benign. Hematite mainly serves as the oxygen evolution anode since valence band edge energy level lies at 1.6 V vs SCE at pH 14 and its CB lies below the hydrogen evolution potential [33]. Theoretically, it shows up to 16% of maximum solar-to-hydrogen conversion efficiency when biased by a photovoltaic device absorbing light below 2 eV. Unfortunately, due to several reasons such as fast recombination and trapping, including low oxygen evolution kinetics and short diffusion lengths, pure hematite is not efficient to cause economical water oxidation. However, its limitations can be overcome by tailoring its structure. Several species of dopant materials such as Ca^{2+}, Cu^{2+}, Mg^{2+}, Si^{4+}, Zn^{2+}, Ti^{4+}, Ge^{4+}, V^{5+}, Pt^{4+} and Nb^{5+} have been attempted to incorporate into hematite structure which have shown significant progress towards improving its water oxidation performance as they allow collection of photogenerated charge carriers and thus more efficient transport [34–40]. Also, various synthesis methods have also been adopted to generate and optimize films with a range of morphologies [41]. Hu et al. demonstrated the use of Pt as dopant material in Fe_2O_3 to enhance its photoelectrochemical property. Significant variations can be caused in the electronic properties of the material via doping. For e.g. doping hematite with 0.2% Mg^{2+} decreases resistivity with more than 3 orders of magnitude. To enhance the PEC performance, Si has been reported as the most effective dopant. Improvement in the charge transport properties and the overall PEC performance has been shown by Ti [42]. Hahn et. al. demonstrated the chemical, physical and photoelectrochemical properties of Ti or Sn doped α-Fe_2O_3 with increased photoconversion efficiency due to improved electron transport within the bulk of the film and the recombination suppression at the interface of electrolyte-film due to stronger electric field near the surface [43]. Composite of α-Fe_2O_3 also with other materials also promises a potential strategy to enhance PEC water splitting efficiency. For instance, Hou and his group reported the synthesis, design and characterization analysis of α-Fe_2O_3/graphene/$BiV_{1-x}Mo_xO_4$ core-shell nanorod heterojunction array for PEC water splitting. Under Xe lamp irradiation, increased photocurrent density of $\sim$1.97 mA/cm^2 at a potential of 1.0 V vs Ag/AgCl with a high photoconversion efficiency of $\sim$0.53% at potential of -0.04 V vs Ag/AgCl was observed for the heterojunction. "Window effect" between $BiV_{1-x}Mo_xO_4$ shells and α-Fe_2O_3 cores was attributed for pronounced photoelectrochemical property and improved photogenerated charge carriers separation at the interfaces of α-Fe_2O_3 nanorod/graphene/$BiV_{1-x}Mo_xO_4$ [44]. Table 4 summarizes various doped/undoped/composite Fe_2O_3 materials prepared by different synthesis methods to compare their PEC activities.

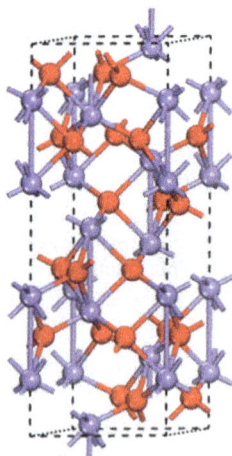

Figure 2: Crystal structure of α-Fe₂O₃ [58].

Table 4. Summary of the Photocurrent Performance for Undoped/doped/composite Fe$_2$O$_3$ Electrodes Reported in Literature

Material	Authors/year	Deposition method	Photocurrent density	Potential	Electrolyte	Illumination source
α-Fe$_2$O$_3$	Tahir et al. [33]	AACVD	600 μA/cm^2	1.23 V vs RHE	1 M NaOH	100 mW/cm^2, AM 1.5
Co-Pi/α-Fe$_2$O$_3$	Zhong et al. [64]	Electrodeposition	~1.2 mA/cm^2	+1.7 V vs RHE	1 M NaOH	150 W Xe arc lamp
n-Fe$_2$O$_3$	Khan et al. [65]	Spray-pyrolytic method	3.7 mA/cm^2		1 M NaOH	50 mW/cm^2 Xenon lamp
Pt-doped α-Fe$_2$O$_3$	Hu et al. [66]	Coelectrode-position	1.43 mA/cm^2	0.4 V vs Ag/AgCl	1 M NaOH	410 mW/cm^2
Si doped α-Fe$_2$O$_3$	Chemelewski et al. [41]	Reactive ballistic deposition	-	-	1 M NaOH	100 mW/cm^2, AM 1.5
Ti- and Sn-doped α-Fe$_2$O$_3$	Hahn et al. [43]	Co-evaporation	0.94 and 0.63 mA/cm^2	1.4 V vs RHE	1M KOH	100 W Xenon lamp
α-Fe$_2$O$_3$/graphene/ BIV$_{1-x}$Mo$_x$O$_4$	Hou et al. [44]	Hydrothermal deposition	~1.97 mA/cm^2	1.0 V vs Ag/AgCl	0.01 M Na$_2$SO$_4$	150 W Xenon lamp

2.4   BiOI

Bismuth iodide, BiOI, a ternary compound, belongs to the Sillen family expressed by [M$_2$O$_2$][I$_m$] (m=2). Bismuth oxyhalides have garnered special attention as a narrow bandgap (~1.8 eV) materials for PEC applications [45]. Fig 3 and Table 5 present the crystal structure model and fundamental properties of BiOI, respectively. The valence band energy declines with rising atomic number of X in BiOX (Cl, Br, I). In visible light, BiOI has the strongest absorption since it has the smallest bandgap among them. When exposed to light, BiOI exhibits remarkable PEC properties due to intrinsically layered structure formed of [Bi$_2$O$_2$]2þ slabs interleaved with double layers of halide atoms upon crystallization [46]. Internal electric fields formed between the slabs are believed to be beneficial for improving the photoactivity of the material by inducing the effective separation of photogenerated e-h pairs. Up to now, different morphologies of BiOI, such as microflowers [47], nanoplatelets [48], nanosheets [49] as well as 3D hierarchical structures [49] have been successfully synthesized by a variety of methods to show their PEC activity. Xie et al. demonstrated a simple, template-free, low temperature and rapid synthesis method to produce uniform BiOIs. Generally accepted preparation methods such as solution-based methods, phase-transfer assisted reactions, precipitation and heat treatment methods may suffer from disadvantages like high pressures, heat treatment, surfactants and additives use, and the need to dispose-off organic solvents [50]. Furthermore, hierarchical bismuth oxyiodide microspheres assembled by nanosheets have been evaluated as good adsorbent materials for eliminating phosphate from aqueous solution for PEC and for photocatalysis applications which are considered as green and cost-effective techniques. Hierarchical nanounits are advantageous in many ways to pronounce the photoactivity of the material. They increase the multireflection of light on the surface and capture more light energy. Abundant interspaces are formed among adjacent nanounits which facilitate the reactants transportation to the surfaces of the material [51]. It is also of substantial interest to extend visible light absorption for wide bandgap materials like TiO_2 and ZnO by coupling with narrow bandgap materials. Multi-dimensional morphological structures of BiOI have also known to play vital roles in improving their photoresponse. They offer many advantages over their bulk counterparts such as anti-aggregation ability, high surface-to-volume ratio, abundant transport paths for small organic molecules, and good recyclability and easy separation. Mi et al. evaluated the performance of two dimensional (2D) square-like BiOI nanosheets which revealed improved results [52]. Also, Xiao reported the facile fabrication of three dimensional (3D) BiOI microspheres which showed enhanced photoabsorption due to high BET surface area, unique energy band structure, good disperse property and high surface-to-volume ratios [53]. Other BiOI PEC heterojunction examples include $BiFeWO_6$/BiOI, BiOI/Bi_2S_3 and BiOI/BiOBr etc [54–56]. Table 6 compares various doped/undoped/composite BiOI materials prepared by different synthesis methods to compare their PEC activities.

Table 5. Fundamental properties of BiOI

Band gap type	Indirect
Band gap (eV)	1.8
CB (eV)	0.57
VB (eV)	2.36
Lattice parameters	a = b = 3.9738 Å, c = 9.3722 Å
Bond length (Å)	Bi-O 2.3343 Bi-I 3.3515

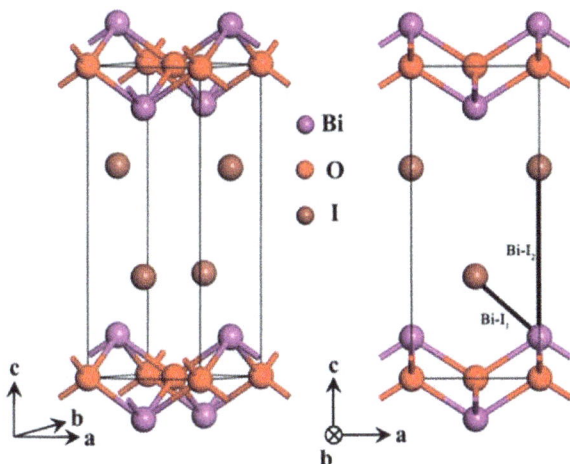

Figure 3: Crystal structure model of BiOI [59].

Table 6. Summary of the Photocurrent Performance for Undoped/doped/composite BiOI Electrodes Reported in Literature

Material	Authors/ year	Deposition method	Photocurrent density	Potential	Electrolyte	Illumination source
BiOI	Hahn et al. [48]	Spray Pyrolysis	0.9 mA/cm^2	0.4 V vs Ag/ AgCl	0.25 M NaI/acetonitrile	100 mW/cm^2
BiOI/BiOBr	Lin et al. [56]	Anion-exchange method	0.035 µA/cm^2	-	0.1M Na$_2$HPO$_4$ + 0.1M NaH$_2$PO$_4$	500W Xe lamp
BiOI/TiO$_2$	Wang et al. [67]	Solvothermal /hydrothermal	0.13 mA/ cm^2	0.46 V vs. Ag/AgCl	0.1 M Na$_2$SO$_4$	300 W Xenon lamp
BiOI	Ge et al. [68]	Solvothermal	-340 µA/cm^2	-	3-Methoxylpropio nitrile	100 mW/cm^2 ,1.5 AM
Graphene-BiOI	Huang et al. [69]	Crystallization	0.9 µA/cm^2	0.0 V. A 5 mV	0.1 M Na$_2$SO$_4$	1 mW/cm^2

Conclusion

The process of splitting water into hydrogen and oxygen has been promising and challenging too, at the same time. This chapter presents the narrow bandgap materials as potential candidates to fit into this category of application. Photoelectrochemical cell is a prominent technique to achieve water splitting to form hydrogen and oxygen. Substantial production of hydrogen at a large scale will tend to solve many commercial problems. Materials with high incident to photon conversion efficiency are required for this purpose. Since narrow band gap materials can harvest more light by utilizing solar spectrum efficiently to behave as photoactive materials, compared to wide bandgap materials. Therefore, this demands a literature survey of well-known narrow band gap materials. Materials like BiOI, CuO, Bi_2S_3 and Fe_2O_3 have been reviewed in detail in this chapter to discuss about their photoactive properties, challenges and ways to improve their efficiency. These materials, in pure form, have shown to be synthesized by various methods, with various morphologies and different dimensions to yield substantial and remarkable photoelectrochemical properties. However, they suffer from certain challenges. To overcome this, their high photoabsorption properties can be further benefited by coupling them with wideband gap materials which themselves are not very capable of absorbing high amount of solar energy. Coupling in the form of composite, heterostructures, co-catalysts and doping exhibit synergistic effects of the system and prove beneficial to both the counterparts. Narrow bandgap materials are a strong absorber for PEC and photocatalysis application and can be exploited further for future applications. Continuous developments and improvements in terms of durability, efficiency and cost are still needed for market viability. Ongoing progress in research and evolution of PEC materials, devices, and systems is making important strides in the field of photoresponsive and PEC applications to improve efficiency through enhanced sunlight absorption.

References

[1]   S. Singh, S. Jain, V. PS, A.K. Tiwari, M.R. Nouni, J.K. Pandey, S. Goel, Hydrogen: A sustainable fuel for future of the transport sector, Renew. Sustain. Energy Rev. 51 (2015) 623–633. https://doi.org/10.1016/j.rser.2015.06.040

[2]   N.S.A. Case, B.S. Bi, Z. Zhang, W. Wang, L. Wang, S. Sun, Enhancement of visible-light photocatalysis by coupling with narrow-band-gap semiconductor: A Case Study on Bi_2S_3/Bi_2WO_6, (2012) 593–597. https://doi.org/10.1021/am2017199

[3]   R. Suarez, P.K. Nair, P. V Kamat, Photoelectrochemical behavior of Bi2S3 nanoclusters and nanostructured thin films, Langmuir 12 (1998) 3236–3241. https://doi.org/10.1021/la9801662

[4]   A.A. Tahir, M.A. Ehsan, M. Mazhar, K.G.U. Wijayantha, M. Zeller, A.D. Hunter, Photoelectrochemical and photoresponsive properties of Bi_2S_3 nanotube and nanoparticle thin films, (2010) 5084–5092. https://doi.org/10.1021/cm101642b

[5]   Y. Bessekhouad, D. Robert, J. V Weber, Bi_2S_3/TiO_2 and CdS/TiO_2 heterojunctions as an available configuration for photocatalytic degradation of organic pollutant, J. Photochem. Photobiol. A Chem. 163 (2004) 569–580. https://doi.org/10.1016/j.jphotochem.2004.02.006

[6]   A. Jana, C. Bhattacharya, J. Datta, Enhanced photoelectrochemical activity of electro-synthesized $CdS–Bi_2S_3$ composite films grown with self-designed cross-linked structure, Electrochim. Acta 55 (2010) 6553–6562. https://doi.org/10.1016/j.electacta.2010.06.022

[7]   X. Lu, F. Pu, Y. Xia, W. Huang, Z. Li, Facile fabrication of porous thin films of $Bi2O_3/Bi_2S_3$ nanocomposite semiconductors at gas/liquid interface and their photoelectrochemical performances, Appl. Surf. Sci. 299 (2014) 131–135. https://doi.org/10.1016/j.apsusc.2014.01.196

[8]   X. Gao, H. Bin Wu, L. Zheng, Y. Zhong, Y. Hu, X.W. Lou, Formation of mesoporous heterostructured $BiVO_4/Bi_2S_3$ hollow discoids with enhanced photoactivity, Angew. Chemie Int. Ed. 126 (2014) 6027–6031. https://doi.org/10.1002/ange.201403611

[9]   C. Liu, Y. Yang, W. Li, J. Li, Y. Li, Q. Chen, Construction of novel Bi_2S_3 nanobelt@WO_3 nanoplate arrays on FTO glass with high photoelectrochemical activity, Int. J. Hydrogen Energy 41 (2016) 5878–5886. https://doi.org/10.1016/j.ijhydene.2016.01.171

[10]   G. Ai, R. Mo, Q. Chen, H. Xu, S. Yang, H. Li, J. Zhong, TiO_2/Bi_2S_3 core–shell nanowire arrays for photoelectrochemical hydrogen generation, RSC Adv. 5 (2015) 13544–13549. https://doi.org/10.1039/C4RA15820B

[11]   Y. Wan, M. Han, L. Yu, J. Jia, G. Yi, Fabrication and photoelectrochemical properties of $TiO_2/CuInS_2/Bi_2S_3$ core/shell/shell nanorods electrodes, RSC Adv. 5 (2015) 78902–78909. https://doi.org/10.1039/C5RA14548A

[12]   F.-A. Liu, Y.-C. Yang, J. Liu, W. Huang, Z.-L. Li, Preparation of $Bi_2O_3@$ Bi_2S_3 core–shell nanoparticle assembled thin films and their photoelectrochemical and

photoresponsive properties, J. Electroanal. Chem. 665 (2012) 58–62.
https://doi.org/10.1016/j.jelechem.2011.11.015

[13]   X. Rong, F. Qiu, J. Yan, H. Zhao, X. Zhu, D. Yang, Coupling with a narrow-band-
gap semiconductor for enhancement of visible-light photocatalytic activity:
preparation of $Bi_2S_3/g-C_3N_4$ and application for degradation of RhB, RSC Adv. 5
(2015) 24944-24952. https://doi.org/10.1039/C4RA15715J

[14]   O.K. Okoth, K. Yan, Y. Liu, J. Zhang, Graphene-doped Bi_2S_3 nanorods as visible-
light photoelectrochemical aptasensing platform for sulfadimethoxine detection,
Biosens. Bioelectron. 86 (2016) 636–642. https://doi.org/10.1016/j.bios.2016.07.037

[15]   D. Chauhan, V.R. Satsangi, S. Dass, R. Shrivastav, Preparation and
characterization of nanostructured CuO thin films for photoelectrochemical splitting of
water, Bull. Mater. Sci. 29 (2006) 709–716.

[16]   G. Wang, H. Wang, Y. Ling, Y. Tang, X. Yang, R.C. Fitzmorris, C. Wang, J.Z.
Zhang, Y. Li, Hydrogen-treated TiO_2 nanowire arrays for photoelectrochemical water
splitting, Nano Lett. 11 (2011) 3026–3033. https://doi.org/10.1021/nl201766h

[17]   Y. Ling, G. Wang, J. Reddy, C. Wang, J.Z. Zhang, Y. Li, The influence of oxygen
content on the thermal activation of hematite nanowires, Angew. Chemie Int. Ed. 124
(2012) 4150–4155. https://doi.org/10.1002/ange.201107467

[18]   M.P. Dare-Edwards, J.B. Goodenough, A. Hamnett, P.R. Trevellick,
Electrochemistry and photoelectrochemistry of iron (III) oxide, J. Chem. Soc. Faraday
Trans. 1 Phys. Chem. Condens. Phases. 79 (1983) 2027–2041.
https://doi.org/10.1039/f19837902027

[19]   P. Salvador, Hole diffusion length in n-TiO_2 single crystals and sintered
electrodes: Photoelectrochemical determination and comparative analysis, J. Appl.
Phys. 55 (1984) 2977–2985. https://doi.org/10.1063/1.333358

[20]   E. Hendry, M. Koeberg, B. O'regan, M. Bonn, Local field effects on electron
transport in nanostructured TiO_2 revealed by terahertz spectroscopy, Nano Lett. 6
(2006) 755–759. https://doi.org/10.1021/nl0600225

[21]   A. Kargar, Y. Jing, S.J. Kim, C.T. Riley, X. Pan, D. Wang, ZnO/CuO
Heterojunction branched nanowires for photoelectrochemical hydrogen generation,
ACS Nano (2013) 11112–11120. https://doi.org/10.1021/nn404838n

[22]   I.S. Cho, Z. Chen, A.J. Forman, D.R. Kim, P.M. Rao, T.F. Jaramillo, X. Zheng,
Branched TiO_2 nanorods for photoelectrochemical hydrogen production, Nano Lett. 11
(2011) 4978–4984. https://doi.org/10.1021/nl2029392

[23]   X. Yang, A. Wolcott, G. Wang, A. Sobo, R.C. Fitzmorris, F. Qian, J.Z. Zhang, Y. Li, Nitrogen-doped ZnO nanowire arrays for photoelectrochemical water splitting, Nano Lett. 9 (2009) 2331–2336. https://doi.org/10.1021/nl900772q

[24]   W. Siripala, A. Ivanovskaya, T.F. Jaramillo, S.-H. Baeck, E.W. McFarland, A Cu_2O/TiO_2 heterojunction thin film cathode for photoelectrocatalysis, Sol. Energy Mater. Sol. Cells. 77 (2003) 229–237. https://doi.org/10.1016/S0927-0248(02)00343-4

[25]   X. Guo, P. Diao, D. Xu, S. Huang, Y. Yang, T. Jin, Q. Wu, M. Xiang, M. Zhang, ScienceDirect CuO/Pd composite photocathodes for photoelectrochemical hydrogen evolution reaction, (2014) 1–11. https://doi.org/10.1016/j.ijhydene.2014.03.084

[26]   F.P. Koffyberg, F.A. Benko, A photoelectrochemical determination of the position of the conduction and valence band edges of p-type CuO, J. Appl. Phys. 53 (1982) 1173–1177. https://doi.org/10.1063/1.330567

[27]   H.-J. Choi, M. Kang, Hydrogen production from methanol/water decomposition in a liquid photosystem using the anatase structure of Cu loaded TiO_2, Int. J. Hydrogen Energy 32 (2007) 3841–3848. https://doi.org/10.1016/j.ijhydene.2007.05.011

[28]   C. Chiang, Y. Shin, S. Ehrman, Li Doped CuO Film Electrodes for Photoelectrochemical Cells, J. Electrochem. Soc. 159 (2012) 227–231. https://doi.org/10.1149/2.081202jes

[29]   S. Choudhary, A. Solanki, S. Upadhyay, N. Singh, V.R. Satsangi, R. Shrivastav, S. Dass, Nanostructured $CuO/SrTiO_3$ bilayered thin films for photoelectrochemical water splitting, J. Solid State Electrochem. 17 (2013) 2531–2538. https://doi.org/10.1007/s10008-013-2139-7

[30]   H.S. Park, C.-Y. Lee, E. Reisner, Photoelectrochemical reduction of aqueous protons with a $CuO|CuBi_2O_4$ heterojunction under visible light irradiation, Phys. Chem. Chem. Phys. 16 (2014) 22462–22465. https://doi.org/10.1039/C4CP03883E

[31]   J.Y. Zheng, G. Song, C.W. Kim, Y.S. Kang, Facile preparation of p-CuO and p-CuO/n-$CuWO_4$ junction thin films and their photoelectrochemical properties, Electrochim. Acta 69 (2012) 340–344. https://doi.org/10.1016/j.electacta.2012.03.011

[32] A.A. Dubale, C.J. Pan, A.G.Tamirat, H.M. Chen, W.N.Su, C.H. Chen, J. Rick, D.W. Ayele, B.A. Aragaw, J.F. Lee, Y.W.Yang, B.J. Hwang, Heterostructured Cu_2O/CuO decorated with nickel as a highly efficient photocathode for photoelectrochemical water reduction, J. Mater. Chem. A 3 (2015) 12482-12499. https://doi.org/10.1039/C5TA01961C

[33]   A.A. Tahir, K.G. Upul Wijayantha, S. Saremi-Yarahmadi, M. Maznar, V. McKee, Nanostructured α-Fe$_2$O$_3$ thin films for photoelectrochemical hydrogen generation, Chem. Mater. 21 (2009) 3763–3772. https://doi.org/10.1021/cm803510v

[34]   V.M. Aroutiounian, V.M. Arakelyan, G.E. Shahnazaryan, G.M. Stepanyan, E.A. Khachaturyan, H. Wang, J.A. Turner, Photoelectrochemistry of semiconductorelectrodes made of solid solutions in the system Fe$_2$O$_3$–Nb$_2$O$_5$, Sol. Energy. 80 (2006) 1098–1111. https://doi.org/10.1016/j.solener.2005.10.005

[35]   A.B. Murphy, P.R.F. Barnes, L.K. Randeniya, I.C. Plumb, I.E. Grey, M.D. Horne, J.A. Glasscock, Efficiency of solar water splitting using semiconductor electrodes, Int. J. Hydrogen Energy. 31 (2006) 1999–2017. https://doi.org/10.1016/j.ijhydene.2006.01.014

[36]   T. Arai, Y. Konishi, Y. Iwasaki, H. Sugihara, K. Sayama, High-throughput screening using porous photoelectrode for the development of visible-light-responsive semiconductors, J. Comb. Chem. 9 (2007) 574–581. https://doi.org/10.1021/cc0700142

[37]   R. Shinar, J.H. Kennedy, Photoactivity of doped α-Fe$_2$O$_3$ electrodes, Sol. Energy Mater. 6 (1982) 323–335. https://doi.org/10.1016/0165-1633(82)90038-7

[38]   C. Leygraf, M. Hendewerk, G.A. Somorjai, Photodissociation of water by p-and n-type polycrystalline iron oxides by using visible light (≤ 2.7 eV) in the absence of external potential, Proc. Natl. Acad. Sci. 79 (1982) 5739–5741. https://doi.org/10.1073/pnas.79.18.5739

[39]   U. Bjoerksten, J. Moser, M. Graetzel, Photoelectrochemical studies on nanocrystalline hematite films, Chem. Mater. 6 (1994) 858–863. https://doi.org/10.1021/cm00042a026

[40]   C. Jorand Sartoretti, B.D. Alexander, R. Solarska, I.A. Rutkowska, J. Augustynski, R. Cerny, Photoelectrochemical oxidation of water at transparent ferric oxide film electrodes, J. Phys. Chem. B. 109 (2005) 13685–13692. https://doi.org/10.1021/jp051546g

[41]   W.D. Chemelewski, N.T. Hahn, C.B. Mullins, Effect of Si doping and porosity on hematite's (α-Fe$_2$O$_3$) photoelectrochemical water oxidation performance, J. Phys. Chem. C. 116 (2012) 5255–5261. https://doi.org/10.1021/jp210877u

[42]   Y.-S. Hu, A. Kleiman-Shwarsctein, A.J. Forman, D. Hazen, J.-N. Park, E.W. McFarland, Pt-doped α-Fe$_2$O$_3$ thin films active for photoelectrochemical water splitting, Chem. Mater. 20 (2008) 3803–3805. https://doi.org/10.1021/cm800144q

[43]   N.T. Hahn, C.B. Mullins, Photoelectrochemical performance of nanostructured Ti- and Sn-doped α-Fe$_2$O$_3$ photoanodes, Chem. Mater. 22 (2010) 6474–6482. https://doi.org/10.1021/cm1026078

[44]   Y. Hou, F. Zuo, A. Dagg, P. Feng, Visible Light-Driven α-Fe$_2$O$_3$ Nanorod/Graphene/BiV$_{1-x}$Mo$_x$O$_4$ Core/Shell Heterojunction Array for Efficient Photoelectrochemical Water Splitting, Nano Lett. 12 (2012) 6464-6473. https://doi.org/10.1021/nl303961c

[45]   Y. Lei, G. Wang, S. Song, W. Fan, M. Pang, J. Tang, H. Zhang, Room temperature, template-free synthesis of BiOI hierarchical structures: visible-light photocatalytic and electrochemical hydrogen storage properties, Dalton Trans. 39 (2010) 3273–3278. https://doi.org/10.1039/b922126c

[46]   S. Singh, R. Sharma, M. Khanuja, A review and recent developments on strategies to improve the photocatalytic elimination of organic dye pollutants by BiOX (X = Cl , Br , I , F) nanostructures, Korean J. Chem. Eng. 35 (2018) 1955-1968. https://doi.org/10.1007/s11814-018-0112-y

[47]   H. Lin, C. Zhou, J. Cao, S. Chen, Ethylene glycol-assisted synthesis, photoelectrochemical and photocatalytic properties of BiOI microflowers, Chinese Sci. Bull. 59 (2014) 3420–3426. https://doi.org/10.1007/s11434-014-0433-0

[48]   N.T. Hahn, S. Hoang, J.L. Self, C.B. Mullins, Spray pyrolysis deposition and photoelectrochemical properties of n-Type BiOI nanoplatelet thin films, ACS Nano 6 (2012) 7712–7722. doi:10.1021/nn3031063. https://doi.org/10.1021/nn3031063

[49]   L. Shan, L. He, J. Suriyaprakash, L. Yang, Photoelectrochemical (PEC) water splitting of BiOI {001} nanosheets synthesized by a simple chemical transformation, J. Alloys Compd. 665 (2016) 158–164. https://doi.org/10.1016/j.jallcom.2016.01.008

[50]   S. Xie, K. Ouyang, X. Ma, Low temperature synthesis of plate-like BiOIs and their highly enhanced visible light photocatalytic performance, Ceram. Int. 40 (2014) 12353–12357. https://doi.org/10.1016/j.ceramint.2014.04.081

[51]   Q.C. Liu, D.K. Ma, Y.Y. Hu, Y.W. Zeng, S.M. Huang, Various bismuth oxyiodide hierarchical architectures: Alcohothermal- controlled synthesis, photocatalytic activities, and adsorption capabilities for phosphate in water, ACS Appl. Mater. Interfaces 5 (2013) 11927–11934. https://doi.org/10.1021/am4036702

[52]   Y. Mi, M. Zhou, L. Wen, H. Zhao, Y. Lei, 6-A highly efficient visible-light driven photocatalyst: two dimensional square-like bismuth oxyiodine nanosheets, Dalton Trans. 43 (2014) 9549–9556. https://doi.org/10.1039/C4DT00798K

[53]   X. Xiao, W.-D. Zhang, Facile synthesis of nanostructured BiOI microspheres with high visible light-induced photocatalytic activity, J. Mater. Chem. 20 (2010) 5866–5870. https://doi.org/10.1039/c0jm00333f

[54]   M. Fang, H. Jia, W. He, Y. Lei, L. Zhang, Z. Zheng, Construction of flexible photoelectrochemical solar cells based on ordered nanostructural $BiOI/Bi_2S_3$ heterojunction films, Phys. Chem. Chem. Phys. 17 (2015) 13531–13538. https://doi.org/10.1039/C4CP05749J

[55]   A. Malathi, P. Arunachalam, A.N. Grace, J. Madhavan, A.M. Al-Mayouf, A robust visible-light driven $BiFeWO_6/BiOI$ nanohybrid with efficient photocatalytic and photoelectrochemical performance, Appl. Surf. Sci. 412 (2017) 85–95. https://doi.org/10.1016/j.apsusc.2017.03.199

[56]   H. Lin, H. Ye, X. Li, J. Cao, S. Chen, Facile anion-exchange synthesis of BiOI/BiOBr composite with enhanced photoelectrochemical and photocatalytic properties, Ceram. Int. 40 (2014) 9743–9750. https://doi.org/10.1016/j.ceramint.2014.02.060

[57]   R.S. Silva, H.D. Mikhail, N.F. Cano, N.O. Dantas, Nanocrystals embedded in a glass matrix, Molecules 22 (2017) 1142. https://doi.org/10.3390/molecules22071142

[58]   B. Wang, Z. Song, B. Wang, J. Yu, C. Ma, C. Zhou, T. Chen, Density functional study on the heterogeneous oxidation of NO over $\alpha\text{-}Fe_2O_3$ catalyst by H_2O_2: Effect of oxygen vacancy, Appl. Surf. Sci. 413 (2017) 292–301. https://doi.org/10.1016/j.apsusc.2017.04.011

[59]   W. Dai, Z. Zhao, Electronic Structure and Optical Properties of BiOI as a Photocatalyst Driven by Visible Light, Inorg. Chem. 54 (2016) 10732–10737. https://doi.org/10.1021/acs.inorgchem.5b01714

[60]   R.S. Mane, B.R. Sankapal, C.D. Lokhande, Photoelectrochemical cells based on chemically deposited nanocrystalline Bi_2S_3 thin films, Mater. Chem. Phys. 60 (1999) 196–203. https://doi.org/10.1016/S0254-0584(99)00085-1

[61]   R. Brahimi, Y. Bessekhouad, A. Bouguelia, M. Trari, Visible light induced hydrogen evolution over the heterosystem Bi_2S_3/TiO_2, Catal. Today. 122 (2007) 62–65. https://doi.org/10.1016/j.cattod.2007.01.030

[62]   Q. Zeng, J. Bai, J. Li, Y. Li, X. Li, B. Zhou, Combined nanostructured Bi_2S_3/TNA photoanode and Pt/SiPVC photocathode for efficient self-biasing photoelectrochemical hydrogen and electricity generation, Nano Energy. 9 (2014) 152–160. https://doi.org/10.1016/j.nanoen.2014.06.023

[63]   K.M. Gadave, C.D. Lokhande, P.P. Hankare, Characterization of CdS-Bi$_2$S$_3$
pellets prepared by co-precipitation method, Mater. Chem. Phys. 38 (1994) 393–397.
https://doi.org/10.1016/0254-0584(94)90219-4

[64]   D.K. Zhong, D.R. Gamelin, Photoelectrochemical water oxidation by cobalt
catalyst (" Co - Pi ")/α-Fe$_2$O$_3$ Composite Photoanodes : Oxygen Evolution and
resolution of a kinetic bottleneck, J. Am. Chem. Soc.132 (2010) 4202–4207.
https://doi.org/10.1021/ja908730h

[65]   S.U.M. Khan, J. Akikusa, Photoelectrochemical splitting of water at
nanocrystalline n-Fe$_2$O$_3$ thin-film electrodes, J. Phys. Chem. B. 34 (1999) 7184–7189.
https://doi.org/10.1021/jp990066k

[66]   Y. Hu, A. Kleiman-shwarsctein, A.J. Forman, D. Hazen, J. Park, E.W. Mcfarland,
S. Barbara, S. Barbara, Pt-Doped α-Fe$_2$O$_3$ thin films active for photoelectrochemical
water splitting, Chem. Mater. 20 (2008) 3803–3805.
https://doi.org/10.1021/cm800144q

[67]   L. Wang, W.A. Daoud, BiOI/TiO$_2$-nanorod array heterojunction solar cell :
Growth, charge transport kinetics and photoelectrochemical properties, Appl. Surf.
Sci. 324 (2015) 532–537. https://doi.org/10.1016/j.apsusc.2014.10.110

[68]   S. Ge, K. Zhao, L. Zhang, Microstructure-dependent photoelectrochemical and
photocatalytic properties of BiOI, J. Nanoparticle Res. 14 (2012) 1015.
https://doi.org/10.1007/s11051-012-1015-1

[69]   H. Huang, K. Liu, Y. Zhang, K. Chen, Y. Zhang, N. Tian, Tunable 3D hierarchical
graphene–BiOI nanoarchitectures: Their in situ preparation, and highly improved
photocatalytic performance and photoelectrochemical properties under visible light
irradiation, RSC Adv. 4 (2014) 49386–49394. https://doi.org/10.1039/C4RA07533A

Photoelectrochemical Water Splitting: Materials and Applications   Materials Research Forum LLC
Materials Research Foundations **70** (2020) 110-128   https://doi.org/10.21741/9781644900734-5

Chapter 5

Ti-based Materials for Photoelectrochemical Water Splitting

Chin Wei Lai[*], Nur Azimah Abd Samad, Mohd Rafie Johan

[1]Nanotechnology & Catalysis Research Centre (NANOCAT), Level 3, Block A, Institute for Advanced Studies (IAS), University of Malaya (UM), 50603 Kuala Lumpur, Malaysia

*cwlai@um.edu.my

Abstract

One of the most promising prospects for efficient renewable resources is the production of hydrogen (H_2) gas. Extensive research on the development of n-type semiconductors for photoelectrochemical (PEC) water splitting process using solar energy is needed to bring H_2 to the point of commercial readiness and viability in terms of performance and cost. Recent studies on TiO_2 have recommended TiO_2 as the leading candidate for PEC water splitting due to its low cost, non-toxicity, self-cleaning property, ready availability, strong photocatalytic activity, and stability against photocorrosion. In a PEC water splitting cell, the high efficiency of TiO_2 as a photoelectrode requires an appropriate architecture that can minimize the loss of electrons at nanostructure connections and maximize photon absorption. To further improve the immigration of photo-induced charge carriers, considerable efforts are needed to further improve the performance of water splitting under visible light illumination. In the subsequent sections, the historical overview, basic principle, material selection and work done by various researchers with regards to the TiO_2 based materials applied in PEC water splitting will be reviewed in detail.

Keywords

TiO_2, Hydrogen, Photoelectrochemical Water Splitting, Photocatalytic Activity, Visible Light Illumination

Contents

1.   Introduction

As the drive to search for alternative energy sources worldwide is growing significantly, interest in cost effective energy sources that can easily produce and are highly efficient is increasing [1]. One such source is the use of sunlight energy for exciting semiconductor materials, later coined as semiconductor photocatalyst [2-4]. The early work on semiconductor photocatalysis was reported by Becquerel (1839). He reported that an electrode connecting to silver chloride electrode and immersed in an electrolyte solution generated electric current and voltage when being exposed to sunlight illumination. The discovery that the energy of sunlight can be captured and converted into electric power had brought out lots of great ideas for scientists and researchers engaged in search of alternative energy sources. Later, the photoelectric effect was first applied to a device by Charles Fritts (1883) with the development of a selenium and gold *pn* junction device with approximately 1% efficiency. However, the literature regarding to the photocatalyst was limited at the early part of the last century.

In 1954, the first *pn* junction solar cell design was published and reported by Bell Laboratories with an efficiency of 6% [5]. The innovation by Bell Laboratories produced the first viable commercial solar cell, which has been revolutionized the photovoltaic industry. Since then, the improvements have been made to give photovoltaic more accessibility in the global energy market. Sunlight was converted into electricity with solid-state interconnecting devices, which often consist of silicon[6]. But the emergent generation of PEC water splitting cells (photovoltaics integration with an electrolyzer for clean and mobile H$_2$ energy supplies) is challenging this dominance. The main disadvantage of the photovoltaic is that it does not operate well at night or during the period of bad weather. Thus, storage of energy as chemical fuel in the H$_2$ form is essential [3, 7, 8].

The energy stored in H_2 fuel within a fuel cell can be efficiently converted into electrical energy and to be used at all times. This cell generally consists of nanocrystalline materials that generate O_2 and offer an economical production operation together with other attractions like high chemical stability and the flexibility in an aqueous solution. These materials also have high light incidence in PEC water splitting cell [7, 9-11]. In 1972, Fujishima and Honda discovered the PEC water splitting process for H_2 generation using TiO_2 electrodes. This breakthrough has triggered the subsequent interests in photocatalysis research by scientists and researchers from all over the world on TiO_2 and made TiO_2 as an important component in many practical applications [12].

2. Basic principle of PEC water splitting

The PEC water splitting process is the general term for a chemical reaction, in which water is photocatalytical splitted into O_2 and H_2 and the water splitting reaction is being catalyzed. A fundamental principle schematic diagram of such overall water splitting reaction using a semiconductor photocatalyst is presented in Figure 1.

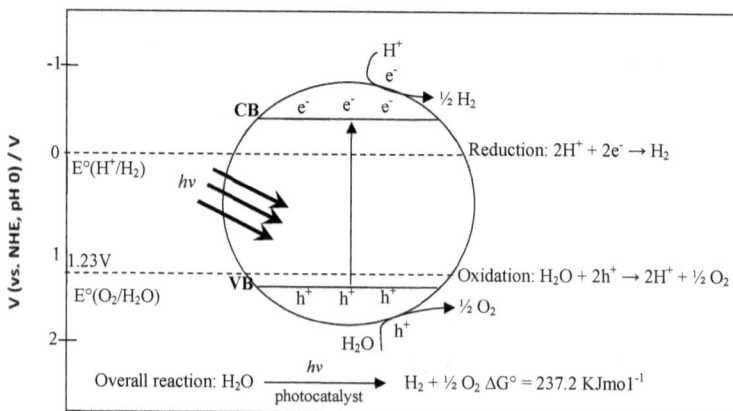

Figure 1: The fundamental principle of overall water splitting reaction on a semiconductor photocatalyst.

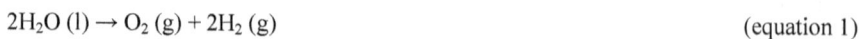

$$2H_2O \, (l) \rightarrow O_2 \, (g) + 2H_2 \, (g) \qquad \text{(equation 1)}$$

The complete reaction of water splitting process is a thermodynamically upward reaction with a large Gibbs free energy of $\Delta G^0 = + 237.2$ KJ mol^{-1} (equation 1). This shows that photon energy is required to overcome the significant positive change in Gibbs free energy by the PEC water [13]. When n-type TiO_2 photocatalyst absorbs photon (hv) with more energy than the energy of their band gap, the light driven water splitting process is triggered. This absorption of light generates electrons in the conduction band and holes in the valence band (equation 2). The holes work on the TiO_2 electrolyte interface by oxidizing water molecules to create O_2 and H^+ ions (equation 3). The electrons move to the platinum electrode via the external circuit where the H^+ ions generate H_2 molecules by electric field or external bias (equation 4).

$$TiO_2 + 2\ hv \rightarrow TiO_2 + 2e^- + 2h^+ \qquad \text{(equation 2)}$$

$$2H_2O + 4h^+ \rightarrow O_2 + 4H^+ \qquad \text{(equation 3)}$$

$$2H^+ + 2e^- \rightarrow H_2 \qquad \text{(equation 4)}$$

If these reactions have to proceed with success, the band gap of the semiconductor photocatalyst must satisfy certain requirements. The photon that supplies energy to the reaction must have sufficient energy, both to oxidize and to reduce the reactions of water on its surface. In PEC water splitting, it implies that the incoming energy must be at least equilibrium with cell potential for water electrolysis, which is 1.23 eV at 25 °C and 1 atm.

In this manner, Fermi level must be constant throughout the sample; otherwise no electric current will flow. A potential gradient along the interface is formed and then will lead to the band bending. Thus, photo-induced electrons will force to transfer from semiconductor to the metal. These photo-induced electrons are then transmitted into the counter electrode and reduced H^+ ions into H_2 under external bias. However, any process of solar conversion involving photocatalyst semiconductor causes inevitable energy losses. When solar light hits the photocatalyst, the electrons from the valence band move to the conduction band. The electrons have been excited by photon absorption which the energy is greater than the photocatalyst band gap. In this case, electron/hole pair recombination can occur very quickly and release energy in the form of unproductive heat and photons [9]. Other possible reasons for the energy loss include transport of electrons in the sample during charge carriers separation, transporting electrons from photoanode to counter electrode, and heat from Joule with electrons flow through the external circuit [13]. The energy losses can be higher than 0.8 eV in a practical PEC water splitting system [14, 15]. Therefore, the optimal band gap in the PEC water splitting cell of semiconductor photocatalyst must be approximately 2.0 eV [9].

However, any solar energy conversion process involving the photocatalyst semiconductor can cause unavoidable energy loss. For an example, when solar illumination hits photocatalysts with absorbed photons are larger than the band gap for photocatalysts, as a result energy is lost as heat. Only partial of the stimulated energy can be used for the production of electron/hole pairs from electrons and atomic bonds [9]. Other possible energy losses include transmission of electrons within the photoanode during the separation of the charge carrier, transporting electrons from photoanode to platinum electrode and Joule heating due to electron flow through the external circuit [13]. In a practical PEC water splitting system, energy losses can be greater than 0.8 eV [14, 15]. Thus, the optimum band gap of semiconductor photocatalyst applied to PEC water splitting cell is approximately 2.0 eV [9].

3.   Material selection for PEC water splitting

Considering the processes involved with dissociating water from particulate photocatalysts under visible light radiation, photocatalyst materials must meet several functional energy and electrochemical properties requirements of the band gap, as shown below:

    i.Band edge position: The level of the conductive band should be more negative than the level of H_2 production ($E_{H2/H2O}$), while the level of the valence band should be more positive than the level of water oxidation ($E_{O2/H2O}$) for efficient H_2 production [14].

    ii.   Band-gap: For most of the photocatalyst, the electronic band gap should be low in order to be used for photoexcitation under solar illumination [14].

    iii. Transportation of charge carriers: Charge carriers should be transported to the counter electrode for highly efficient H_2 generation with minimal losses from the bulk oxide material [9].

    iv. Stability: The photocatalyst must be stable against photocorrosion in electrolyte [9].

4.   TiO$_2$ photocatalyst for PEC water splitting

In 1972, Fujishima and Honda successfully reported the PEC water splitting with TiO_2 as a photoelectrode [12]. TiO_2 has now been widely used in the PEC water splitting system for H_2 generation as an effective photoelectrode due to its unique features [16-19]. In previous studies, particles/spheres of TiO_2 as photoelectrodes were reported as being freely moving in solution suspension or compacted into photoelectrodes in a PEC water splitting system. The drawbacks of this photoelectrode, however, include defects or trapping sites, additional grain boundaries, and disorders between two particle /sphere contact areas [20, 21]. Therefore, in the bulk phase TiO_2 particles/spheres the electron transport time was rather long, which led to more recombination losses and problems of scattering of photo-induced electrons. In addition to this, the use of photoelectrode TiO_2 particles/sphere requires appropriate substrates to support catalysts in the PEC system to ease the photoreactive filtering procedure [22]. TiO_2 photoelectrode, a two-dimensional film, was recently used in PEC water splitting due to its capacity to remove the aforementioned problems and to be still reusable [20, 23]. However, two dimensional TiO_2 photoelectrode thin film does not generally have a large active area in PEC reactions [22].]. It is therefore important to maximize the specific surface of the TiO_2 thin film. The density of the active locations available for photo absorption is considered ideal for 1D TiO_2 nanotube arrays because of the existence of its internal and external surface area of the nanotube [24-26]. The use of 1D TiO_2 nanotubes will improves the performance of a PEC water splitting much more efficient and more effective.

Several approaches such as the sol-gel synthesis [27-29], hydrothermal synthesis [30-32], and the anodization technique [33-36] have been available in order to achieve synthesis of 1D TiO_2 nanostructures. Anodization technique is most feasible among all those because it can produce anodizing oxides that are self-organized in the form of nanotubes, and almost perfectly vertical alignment. Furthermore, the anodization for large-scale industrial production has been used [37, 38]. Most of all, 1D TiO_2 nanotube arrays provide a unidirectional photo-induced transmission of TiO_2 grains in the tube growth direction by a single electrical channel. This could improve the PEC water splitting performance due to low recombination losses on grain boundaries and improve the vertical transport of charge carriers [21]. A controlled synthesis process must be optimized for the production of nanotube array by setting the length, the wall thickness, the pores and the intertube spacing in the synthesis process in order to obtain the correct dimensions and morphologies.

5.   Tuning the photocatalytic of TiO_2 into the visible light region

The development of TiO_2 assemblies' nanostructures has become the most widely studied with promising functional properties. One of the most studied semiconductor photocatalysts in PEC water splitting applications has been 1D TiO_2 among various nano-architectures of TiO_2. However, its poor visible light response and fast recombination rate of charge carriers have been obvious impediment to the widespread use of TiO_2 as a photoelectrode in PEC water splitting cells [13, 14]. TiO_2 is indeed only efficient for photocatalysts within the ultraviolet region ($\lambda < 400$ nm). About 4-5% of our solar energy constitutes UV light. Therefore, the use of our solar energy's visible light is essential to increase the efficiency of water splitting photoconversion [4, 13].

The only way to overcome this disadvantage is to optimize the dimension of TiO_2 nanotubes, as described in the previous section. But, significant efforts have to be made to minimize the recombination losses in charge carrier by integrating the optimum number of anions doping or transition metal-oxide elements into the TiO_2 lattice, thus to extent spectral response of TiO_2 into the visible spectrum. The effects of narrowing band gap will lead the excitation light to expand into a visible area to provide space for the recovery of carriers to be slowed down [4]. In 2004, Komornicki and co-researchers proposed that photocatalysts for binary oxide were promising candidates with enhanced properties for various applications. The best properties of its pure components may be combined to allow photocatalysts to benefit from the modification of the electronic system structure [39]. Higashimoto and co-researchers subsequently suggested, in 2006, that the binary photocatalyst can promote better separation of charge carriers by building up electrons and holes in two separate semiconductor layers with charge recombination suppressors. In the blending oxide systems, the useful features of the individual components can be complementary to simultaneously reduce the disadvantages [40].

In general, many studies have shown that visible light absorption of TiO_2 photocatalyst can be improved and that charge carrier's recombination can be eliminated when connected to another photocatalyst semiconductor [15, 24, 41-43]. Replacing the cationic ions in the crystal lattice can produce impurity levels to better absorption of visible light regions as shown in Figure 2. In this case, TiO_2 nanotubes have been modified to create the interfacial potential gradient with the corresponding energetic position to make transportation and the separation of the charge carrier easier. When an illumination activated TiO_2 photocatalyst that is coupled with another semiconductor, the electrons will be injected with a more negative band (CB) level than the semiconductor in a positive direction, while a more positive valence band hole (VB) will be transferred from the semiconductor to the negative. As a consequence, charge carriers life span will be increased as well as the interfacial charge transfer to water will be significantly increased [13, 43].

However, photocatalysts performance in case of binary oxides strongly depends upon cationic dopant content. The impurity levels created by cationic dopants in the binary oxide photocatalyst are usually discrete, which would appear disadvantageous for the migration of the charge carriers if the content of cationic dopants exceeds the optimum value [24, 38, 44]. Therefore intuitive conjectures are impossible and it is difficult to conduct focused research. In order to develop an efficient visible-light photocatalyst, it is important to fine-tune the content of the cationic substitution into TiO_2 photocatalysts. Table 1 summarizes the details and results of several researchers' for cationic dopants incorporated in TiO_2 photocatalysts.

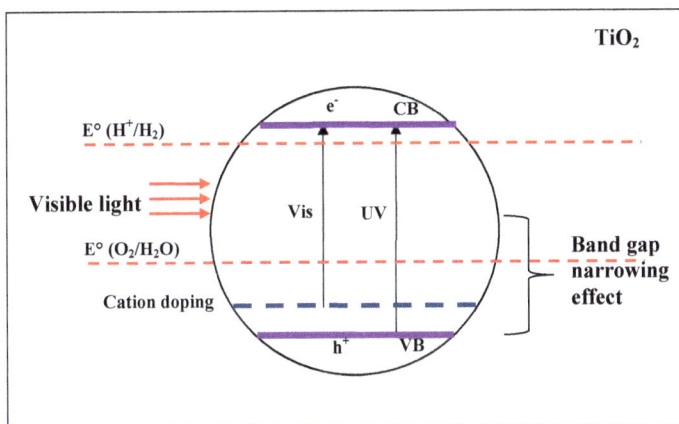

Figure 2: Cationic-loaded band structure with visible light response on the wide band gap of the TiO₂ semiconductor photocatalyst.

From table 1, it is clear that researchers have continuously make progress in investigating and optimizing the content of TiO_2 photocatalyst cationic dopants. Cationic dopants interact with TiO_2 in order to achieve charge neutrality in the consequential system, therefore a binary oxide system is formed. Optimum cationic dopants were found to prolong the life of charge carriers and simultaneously reduce the charge carrier recombination losses. Modified TiO_2 could therefore lead to a better photocatalytic activity than pure TiO_2 and could extent the absorption ability from the UV region into visible region. Cationic-modified TiO_2 acted as an electron receptor by providing an impurity level and, therefore increase the photon absorption under visible range. It plays a big role in photocatalytic activity for binary oxide system.

Table 1: Summary of the past researches on the different cationic dopants incorporated in TiO$_2$ photocatalysts

Cationic dopant	Author	Findings
Cr	Takeuchi *et al.* [46]	In the decomposition of NO under visible light irradiation, Cr-loaded TiO$_2$ was very effective.
Fe	Piera *et al.* [47]	Fe^{3+} ions were used in the TiO$_2$ lattice as shallow traps, with strong photocatalytic characteristics keeping Fe content lower than 2 at%. This is because the charge carriers suffered low recombination losses.
Si	Miyashita *et al* [48]	The TiO$_2$ multi-layer film, which was loaded with Si, showed greater photocatalytic activity and photoconductivity, due to the longer lifetimes of the charge carriers.
Cu	Zhang *et al.* [49]	For best-effective photocatalytic activities, the best Cu-loaded TiO$_2$ was 1.45 at%. The sample's absorption edges shifted to visible region.
V	Wu and Chen [50]	The sol-gel synthesized V-loaded TiO$_2$ photocatalysts provided a promising method for enhancing photocatalytic TiO$_2$ activity under visible light.
Au	Sonawane and Dongare [51]	The shift in the wavelength of the optical absorption due to Au nanoparticles (1-2 at%) was integrated in TiO$_2$ structure. The phenol decomposition rate was improved by two fold with Au-loaded TiO$_2$ photocatalyst as compared to the undoped TiO$_2$ photocatalyst.
Au	Mirescu *et al.* [52]	High selectivity and photocatalytic activity of the glucose oxidation at the reaction temperature at around 40-60 °C were produced in 0.45 wt% of the Au-loaded TiO$_2$.
Ag	Rupa *et al.* [53]	Ag-loaded TiO$_2$ photocatalyst showed better photocatalytic yellow-17 degradation under ultraviolet and visible light radiations because of the minimizing photoinducted charge carriers' recombination loss.
Al	Choi *et al.*) [54]	The Al-loaded TiO$_2$ photocatalyst was applied to a gas sensor that required a high conductivity. It has shown that at a temperature (600-900 °C), the Al-loaded TiO$_2$ is more conductive than the TiO$_2$ pure one.
Cu	Li *et al.* [55]	Cu ions acted effectively in Cu-loaded TiO$_2$ nanotubes, thus supporting photoconversion efficiency. It acted as

		electron trappers prohibiting recombinations of photoexcited electrons/holes.
Ag	Sonaba *et al.* [56]	Ag-loaded TiO_2 was shown to be more efficient with 1.5 at% than TiO_2 itself. It is because of the formation of a Schottky barrier in the contact area of the metal-semiconductor .
Pd	Mohapatra *et al.* [57]	The self-organized TiO_2 nanotube with Pd nanoparticles was an excellent photocatalyst which has been significantly increased by the lifetime of the charge carriers and can decompose non-biodegradable azo dye (methyl red and methyl orange).
Fe	Kuang *et al.* [58]	TiO_2 produced the highest photocurrent with optimum content of 0.5 at% Fe, which was 7 times higher than with the undoped TiO_2 nanotube. The improved PEC behaviors was due to the increased charge carrier separation and the extended the photoresponse of the TiO_2 from UV region and visible region because of the low band gap of 2.2 eV of Fe_2O_3.
Cu	Hou *et al.* [59]	Cu-loaded TiO_2 nanotubes improved efficiency in charge separation and photoconversion.
Zn, Fe	Hou *et al.* [60]	$ZnFe_2O_4$-TiO_2 nanotube arrays revealed corrective behavior and increased separation of photo-induced charge carriers in order to enhancing the capability of photoconversion and photoelectrocatalytic compared to pure TiO_2 nanotube arrays.
Cu	Liu *et al.*, [61]	Cu-loaded TiO_2 nanotubes synthesized with the sonoelectrochemical deposition method showed a photo-electrochemical performance improvement due to the excellent charge carrier separation and transmission characteristics.

In addition to the cationic dopants, the combination of metal oxide-based TiO_2 composites has made endless efforts to improve the photocatalytic activity of TiO_2 under the visible light. For the past decade this new trend has begun. This combination is also known as a heterogeneous material, Incorporating a second metal oxide on top of TiO_2. The second metal oxide for metal oxide composites lies at the grain boundary of TiO_2, creating doping effects involving the change of the concentration of oxygen vacancies, the introduction of free charge carriers transports, and the heterojunction at metal oxides' intergrain [45]. Due to the band alignment between two different metal oxides, the heterojunction existed. Either straddling bandgap (type I) or staggered bandgap (type II) are existed. This heterojunction helps to increase electron transportation from band to band and inhibits recombination losses for the charge carriers, thereby improving photocatalytic activity. Optimal quantity of second metal oxide would lead to better absorption by visible light and reduce recombinant losses for the charge carriers. A summary of past studies for a combination of TiO_2 composites with metal oxides is provided in Table 2.

Table 2: Summary of the past researches on the different metal oxide dopants incorporated in TiO_2 photocatalysts

Second metal oxide	Authors (Year)	Findings
Cu_2O	Lofficial, D. *et al.* [62]	Improved CO_2 conversion to hydrocarbon molecules under sunlight.
GO	Chen,C. *et al.* [63]	Better depolarization of methyl orange under visible light as compared to bare TiO_2.
FeO_x	Tada, H. *et al.* [64]	Improved degradation of 2-naphthol (2-NAP) as compared to P-25 TiO_2 under visible light.
FeO_x	Wu W. *et al.* [65]	About 50% - 60% decomposition of methylene blue (MB) within 90 minutes under Hg light (spectral line of ultraviolet light to visible light) was observed.
Fe_2O_3	Chaguetmi, S. *et al.* [66]	Enhanced photocatalytic activity for methylene blue (MB) photodegradation under mimic sunlight as compared to bare TiO_2.
Co_3O_4	Liu, J. *et al.* [67]	High evolution rates of 41.8 and 22.0 $\mu mol\,h^{-1}\,g^{-1}$ for H_2 and O_2 productions from pure water respectively.
Fe_2O_3	Arifin, K. *et al.* [68]	The highest current density was achieved from 5 minutes deposition of Fe_2O_3 on top of TiO_2 and resulted in wavelength shifted to 518 nm light absorption.
Mn_2O_3	Mansoor, M. A. *et al.* [69]	Improved maximum photocurrent density of 343 $\mu A/cm^2$ at 0.7 V under sunlight.

Conclusion

Since 1972, Fujishima and Honda have discovered water photoelectrolysis with TiO_2 as the photoanode together with the exceptional chemicals and physical properties of TiO_2 have been recognized over the years. The TiO_2 1D nanotubes structure has been heavily considered, particularly in the PEC water splitting application by using the simple electrochemical oxidation reaction of Ti's substrate under certain environmental conditions. However, pure TiO_2 nanotubes are still far from being a potential PEC water splitting system. The bundling issue (nanotube arrays disorder arrangement) and the weak adherence of the Ti substrate nanotube array have remained major challenges. Low visible light absorption and rapid recombination of charge carriers transport systems have also been major issues to limit the widespread use of TiO_2 nanotubes. Therefore, it is challenging to produce a highly efficient PEC water splitting system by the use of TiO_2 nanotubes, unless several problems are addressed. In addition, 1D nanotubular structural features with the desired width, pore diameter and wall width can be used to improve TiO_2 PEC water splitting performance. However, a controlled synthesis procedure for the well-aligned 1D TiO_2 nanotubes needs to be optimized until right measurements and morphologies are reached. TiO_2 with highly ordered nanotubes lead generally to an undesirable problem of bundling which reduces photoconversion efficiencies significantly in the PEC water splitting system. Continuous efforts have been made with the combination of a promising cationic dopant and metal oxide dopant with TiO_2 in order to improve the efficiency for the separation of charge carriers to resolve the above-mentioned problems. Such findings will help to build the fundamental principles modified-TiO_2 nanotubes in the development of the sustainable energy system's H_2 fuel cell.

Acknowledgements

This research work was financially supported by University of Malaya Grant (No. RP045B-17AET), Impact-Oriented Interdisciplinary Research Grant (No. IIRG018A-2019), and Global Collaborative Programme - SATU Joint Research Scheme (No. ST012-2019).

References

[1] W. Kreuter, H. Hofmann, Electrolysis: the important energy transformer in a world of sustainable energy, Int. J. Hydrogen Energy 23 (1998) 661-666. https://doi.org/10.1016/S0360-3199(97)00109-2.

[2] V. Aroutiounian, V. Arakelyan, G. Shahnazaryan, Metal oxide photoelectrodes for hydrogen generation using solar radiation-driven water splitting, Sol. Energy 78 (2005) 581-592. https://doi.org/10.1016/j.solener.2004.02.002.

[3] D. Barreca, G. Carraro, V. Gombac, A. Gasparotto, C. Maccato, P. Fornasiero, E. Tondello, Supported metal oxide nanosystems for hydrogen photogeneration: Quo vadis?, Adv. Funct. Mater. 21 (2011) 2611-2623. https://doi.org/10.1002/adfm.201100242.

[4] H. Tong, S. Ouyang, Y. Bi, N. Umezawa, M. Oshikiri, J. Ye, Nano-photocatalytic materials: possibilities and challenges, Adv. Mater. 24 (2012) 229-251. https://doi.org/10.1002/adma.201102752.

[5] D.M. Chapin, C. Fuller, G. Pearson, A new silicon p-n junction photocell for converting solar radiation into electrical power, J. Appl. Phys. 25 (1954) 676-677. https://doi.org/10.1063/1.1721711.

[6] H. Okamoto, Y. Sugiyama, H. Nakano, Synthesis and modification of silicon nanosheets and other silicon nanomaterials, Chem. A Eur. J. 17 (2011) 9864-9887. https://doi.org/10.1002/chem.201100641.

[7] M. Grätzel, Photoelectrochemical cells, Nature 414 (2001) 338. https://doi.org/10.1038/35104607.

[8] A. Currao, Photoelectrochemical water splitting, CHIMIA Int. J. Chem. 61 (2007) 815-819. https://doi.org/10.2533/chimia.2007.815.

[9] C.A. Grimes, Synthesis and application of highly ordered arrays of TiO_2 nanotubes, J. Mater. Chem. 17 (2007) 1451-1457. https://doi.org/10.1039/B701168G.

[10] N.K. Allam, C.A. Grimes, Effect of cathode material on the morphology and photoelectrochemical properties of vertically oriented TiO_2 nanotube arrays, Sol. Energ. Mater. Sol. Cells 92 (2008) 1468-1475. https://doi.org/10.1016/j.solmat.2008.06.007.

[11] G. Centi, S. Perathoner, The role of nanostructure in improving the performance of electrodes for energy storage and conversion, Eur. J. Inorg. Chem. 2009 (2009) 3851-3878. https://doi.org/10.1002/ejic.200900275.

[12] A. Fujishima, K. Honda, Electrochemical photolysis of water at a semiconductor electrode, Nature 238 (1972) 37. https://doi.org/10.1038/238037a0.

[13] D.Y. Leung, X. Fu, C. Wang, M. Ni, M.K. Leung, X. Wang, X. Fu, Hydrogen production over titania-based photocatalysts, ChemSusChem 3 (2010) 681-694. https://doi.org/10.1002/cssc.201000014.

[14] M. Ni, M.K. Leung, D.Y. Leung, K. Sumathy, A review and recent developments in photocatalytic water-splitting using TiO$_2$ for hydrogen production, Renewable and Sustainable Energy Reviews 11 (2007) 401-425. https://doi.org/10.1016/j.rser.2005.01.009.

[15] R.M. Navarro Yerga, M.C. Alvarez Galvan, F. Del Valle, J.A. Villoria de la Mano, J.L. Fierro, Water splitting on semiconductor catalysts under visible-light irradiation, ChemSusChem 2 (2009) 471-485. https://doi.org/10.1002/cssc.200900018.

[16] M. Kitano, M. Matsuoka, M. Ueshima, M. Anpo, Recent developments in titanium oxide-based photocatalysts, Appl. Catal. A: Gen. 325 (2007) 1-14. https://doi.org/10.1016/j.apcata.2007.03.013.

[17] K. Shankar, G.K. Mor, M. Paulose, O.K. Varghese, C.A. Grimes, Effect of device geometry on the performance of TiO$_2$ nanotube array-organic semiconductor double heterojunction solar cells, J. Non-Cryst. Solids 354 (2008) 2767-2771. https://doi.org/10.1016/j.jnoncrysol.2007.09.070.

[18] R. Dholam, N. Patel, M. Adami, A. Miotello, Physically and chemically synthesized TiO2 composite thin films for hydrogen production by photocatalytic water splitting, Int. J. Hydrogen Energy 33 (2008) 6896-6903. https://doi.org/10.1016/j.ijhydene.2008.08.061.

[19] J. Yan, F. Zhou, TiO 2 nanotubes: structure optimization for solar cells, J. Mater. Chem. 21 (2011) 9406-9418. https://doi.org/10.1039/C1JM10274E.

[20] L. Sun, S. Zhang, X. Sun, X. He, Effect of the geometry of the anodized titania nanotube array on the performance of dye-sensitized solar cells, J. Nanosci. Nanotechnol. 10 (2010) 4551-4561. https://doi.org/10.1166/jnn.2010.1695.

[21] P. Roy, S. Berger, P. Schmuki, TiO2 nanotubes: synthesis and applications, Angew. Chem. Int. Ed. 50 (2011) 2904-2939. https://doi.org/10.1002/anie.201001374.

[22] L. Yang, Y. Xiao, S. Liu, Y. Li, Q. Cai, S. Luo, G. Zeng, Photocatalytic reduction of Cr (VI) on WO$_3$ doped long TiO$_2$ nanotube arrays in the presence of citric acid, Appl. Catal. B: Environ. 94 (2010) 142-149. https://doi.org/10.1016/j.apcatb.2009.11.002.

[23] Z. Liu, Q. Zhang, T. Zhao, J. Zhai, L. Jiang, 3-D vertical arrays of TiO$_2$ nanotubes on Ti meshes: efficient photoanodes for water photoelectrolysis, J. Mater. Chem. 21 (2011) 10354-10358. https://doi.org/10.1039/C1JM11072A.

[24] Y.C. Nah, I. Paramasivam, P. Schmuki, Doped TiO2 and TiO2 nanotubes: synthesis and applications, ChemPhysChem 11 (2010) 2698-2713. https://doi.org/10.1002/cphc.201000276.

[25] Z. Zhang, M.F. Hossain, T. Takahashi, Photoelectrochemical water splitting on highly smooth and ordered TiO_2 nanotube arrays for hydrogen generation, Int. J. Hydrogen Energy 35 (2010) 8528-8535. https://doi.org/10.1016/j.ijhydene.2010.03.032.

[26] S. So, K. Lee, P. Schmuki, Ultrafast growth of highly ordered anodic TiO_2 nanotubes in lactic acid electrolytes, J. Am. Chem. Soc. 134 (2012) 11316-11318. https://doi.org/10.1021/ja301892g.

[27] P. Hoyer, K. Nishio, H. Masuda, Preparation of regularly structured porous metal membranes with two different hole diameters at the two sides, Thin Solid Films 286 (1996) 88-91. https://doi.org/10.1016/S0040-6090(96)08549-5.

[28] X. Liu, J. Yang, L. Wang, X. Yang, L. Lu, X. Wang, An improvement on sol-gel method for preparing ultrafine and crystallized titania powder, Mater. Sci. Eng. A 289 (2000) 241-245. https://doi.org/10.1016/S0921-5093(00)00901-1.

[29] Z.-Y. Yuan, B.-L. Su, Titanium oxide nanotubes, nanofibers and nanowires, Colloids Surf. A 241 (2004) 173-183. https://doi.org/10.1016/j.colsurfa.2004.04.030.

[30] T. Kasuga, Formation of titanium oxide nanotubes using chemical treatments and their characteristic properties, Thin solid films 496 (2006) 141-145. https://doi.org/10.1016/j.tsf.2005.08.341.

[31] T. Kasuga, M. Hiramatsu, A. Hoson, T. Sekino, K. Niihara, Titania nanotubes prepared by chemical processing, Adv. Mater. 11 (1999) 1307-1311. https://doi.org/10.1002/(sici)1521-4095(199910)11:15<1307::aid-adma1307>3.0.co;2-h.

[32] M. Zhang, Z. Jin, J. Zhang, X. Guo, J. Yang, W. Li, X. Wang, Z. Zhang, Effect of annealing temperature on morphology, structure and photocatalytic behavior of nanotubed $H_2Ti_2O_4(OH)_2$, J. Molecular Catal. A: Chem 217 (2004) 203-210. https://doi.org/10.1016/j.molcata.2004.03.032.

[33] G. Mor, O.K. Varghese, M. Paulose, N. Mukherjee, C.A. Grimes, Fabrication of tapered, conical-shaped titania nanotubes, J. Mater. Res. 18 (2003) 2588-2593. https://doi.org/10.1557/JMR.2003.0362.

[34] P. Mishra, P. Shukla, A. Singh, O. Srivastava, Investigation and optimization of nanostructured TiO_2 photoelectrode in regard to hydrogen production through photoelectrochemical process, Int. J. Hydrogen Energy 28 (2003) 1089-1094. https://doi.org/10.1016/S0360-3199(02)00197-0.

[35] G.K. Mor, O.K. Varghese, M. Paulose, K. Shankar, C.A. Grimes, A review on highly ordered, vertically oriented TiO_2 nanotube arrays: Fabrication, material properties, and solar energy applications, Sol. Energ. Mater. and Sol. Cells 90 (2006) 2011-2075. https://doi.org/10.1016/j.solmat.2006.04.007.

[36] M. Paulose, G.K. Mor, O.K. Varghese, K. Shankar, C.A. Grimes, Visible light photoelectrochemical and water-photoelectrolysis properties of titania nanotube arrays, J. Photochem. Photobio. A: Chem.178 (2006) 8-15. https://doi.org/10.1016/j.jphotochem.2005.06.013.

[37] J.M. Macak, P. Schmuki, Anodic growth of self-organized anodic TiO_2 nanotubes in viscous electrolytes, Electrochim. Acta 52 (2006) 1258-1264. https://doi.org/10.1016/j.electacta.2006.07.021.

[38] A. Ghicov, P. Schmuki, Self-ordering electrochemistry: a review on growth and functionality of TiO 2 nanotubes and other self-aligned MO_x structures, Chem. Commun. (2009) 2791-2808. https://doi.org/10.1039/B822726H.

[39] S. Komornicki, M. Radecka, P. Sobaś, Structural, electrical and optical properties of TiO_2–WO_3 polycrystalline ceramics, Mater. Res. Bull. 39 (2004) 2007-2017. https://doi.org/10.1016/j.materresbull.2004.07.017.

[40] S. Higashimoto, M. Sakiyama, M. Azuma, Photoelectrochemical properties of hybrid WO_3/TiO_2 electrode. Effect of structures of WO3 on charge separation behavior, Thin Solid Films 503 (2006) 201-206. https://doi.org/10.1016/j.tsf.2005.11.110.

[41] J.H. Bang, P.V. Kamat, Solar cells by design: photoelectrochemistry of TiO2 nanorod arrays decorated with CdSe, Adv. Funct. Mater. 20 (2010) 1970-1976. https://doi.org/10.1002/adfm.200902234.

[42] J. Zhang, Y. Wu, M. Xing, S.A.K. Leghari, S. Sajjad, Development of modified N doped TiO 2 photocatalyst with metals, nonmetals and metal oxides, Energy Environ. Sci. 3 (2010) 715-726. https://doi.org/10.1039/B927575D.

[43] H.-i. Kim, J. Kim, W. Kim, W. Choi, Enhanced photocatalytic and photoelectrochemical activity in the ternary hybrid of CdS/TiO_2/WO_3 through the cascadal electron transfer, J. Phys. Chem. C 115 (2011) 9797-9805. https://doi.org/10.1021/jp1122823.

[44] K. Maeda, A. Xiong, T. Yoshinaga, T. Ikeda, N. Sakamoto, T. Hisatomi, M. Takashima, D. Lu, M. Kanehara, T. Setoyama, Photocatalytic overall water splitting promoted by two different cocatalysts for hydrogen and oxygen evolution under

visible light, Angew. Chem. Int. Ed. 49 (2010) 4096-4099.
https://doi.org/10.1002/anie.201001259.

[45] G. Korotcenkov, S. Han, B. Cho, Metal oxide nanocomposites: Advantages and shortcomings for application in conductometric gas sensors, Materials Science Forum, Trans. Tech. Publ. (2016) 223-229.
https://doi.org/10.4028/www.scientific.net/MSF.872.223.

[46] M. Takeuchi, H. Yamashita, M. Matsuoka, M. Anpo, T. Hirao, N. Itoh, N. Iwamoto, Photocatalytic decomposition of NO under visible light irradiation on the Cr-ion-implanted TiO_2 thin film photocatalyst, Catal. Lett. 67 (2000) 135-137.
https://doi.org/10.1023/A:1019065521567.

[47] E. Piera, M.I. Tejedor-Tejedor, M.E. Zorn, M.A. Anderson, Relationship concerning the nature and concentration of Fe (III) species on the surface of TiO_2 particles and photocatalytic activity of the catalyst, Appl. Catal. B: Environ. 46 (2003) 671-685.
https://doi.org/10.1016/S0926-3373(03)00288-1.

[48] K. Miyashita, S.-i. Kuroda, S. Tajima, K. Takehira, S. Tobita, H. Kubota, Photoluminescence study of electron–hole recombination dynamics in the vacuum-deposited SiO_2/TiO_2 multilayer film with photo-catalytic activity, Chem. Phys. Lett. 369 (2003) 225-231. https://doi.org/10.1016/S0009-2614(02)02009-2.

[49] X. Zhang, K. Huo, L. Hu, P.K. Chu, Fabrication and photocatalytic activity of nanoporous WO_3 film, Nanosci. Nanotechnol. Lett. 2 (2010) 51-57.
https://doi.org/10.1166/nnl.2010.1052.

[50] J.C.-S. Wu, C.-H. Chen, A visible-light response vanadium-doped titania nanocatalyst by sol–gel method, J. Photochem. Photobio. A: Chem. 163 (2004) 509-515. https://doi.org/10.1016/j.jphotochem.2004.02.007.

[51] R. Sonawane, M. Dongare, Sol–gel synthesis of Au/TiO_2 thin films for photocatalytic degradation of phenol in sunlight, J. Molecular Catal. A: Chem. 243 (2006) 68-76. https://doi.org/10.1016/j.molcata.2005.07.043.

[52] A. Mirescu, H. Berndt, A. Martin, U. Prüße, Long-term stability of a 0.45% Au/TiO_2 catalyst in the selective oxidation of glucose at optimised reaction conditions, Appl. Catal. A: Gen. 317 (2007) 204-209. https://doi.org/10.1016/j.apcata.2006.10.016.

[53] A.V. Rupa, D. Manikandan, D. Divakar, T. Sivakumar, Effect of deposition of Ag on TiO_2 nanoparticles on the photodegradation of Reactive Yellow-17, J. Hazardous Mater. 147 (2007) 906-913. https://doi.org/10.1016/j.jhazmat.2007.01.107.

[54] Y.J. Choi, Z. Seeley, A. Bandyopadhyay, S. Bose, S.A. Akbar, Aluminum-doped TiO2 nano-powders for gas sensors, Sensors and Actuators B: Chem. 124 (2007) 111-117. https://doi.org/10.1016/j.snb.2006.12.005.

[55] H. Li, X. Duan, G. Liu, L. Li, Synthesis and characterization of copper ions surface-doped titanium dioxide nanotubes, Mater. Res. Bull. 43 (2008) 1971-1981. https://doi.org/10.1016/j.materresbull.2007.10.005.

[56] N. Sobana, K. Selvam, M. Swaminathan, Optimization of photocatalytic degradation conditions of Direct Red 23 using nano-Ag doped TiO_2, Sep. Purif. Technol. 62 (2008) 648-653. https://doi.org/10.1016/j.seppur.2008.03.002.

[57] S.K. Mohapatra, N. Kondamudi, S. Banerjee, M. Misra, Functionalization of self-organized TiO2 nanotubes with Pd nanoparticles for photocatalytic decomposition of dyes under solar light illumination, Langmuir 24 (2008) 11276-11281. https://doi.org/10.1021/la801253f.

[58] S. Kuang, L. Yang, S. Luo, Q. Cai, Fabrication, characterization and photoelectrochemical properties of Fe_2O_3 modified TiO_2 nanotube arrays, Appl. Surf. Sci. 255 (2009) 7385-7388. https://doi.org/10.1016/j.apsusc.2009.04.005.

[59] Y. Hou, X.Y. Li, Q. Zhao, X. Quan, G. Chen, Fabrication of Cu_2O/TiO_2 nanotube heterojunction arrays and investigation of its photoelectrochemical behavior, Appl. Phys. Lett. 95 (2009) 093108. https://doi.org/10.1063/1.3224181.

[60] Y. Hou, X.Y. Li, Q.D. Zhao, X. Quan, G.H. Chen, Electrochemical method for synthesis of a $ZnFe_2O_4/TiO_2$ composite nanotube array modified electrode with enhanced photoelectrochemical activity, Adv. Funct. Mater. 20 (2010) 2165-2174. https://doi.org/10.1002/adfm.200902390.

[61] Y. Liu, H. Zhou, J. Li, H. Chen, D. Li, B. Zhou, W. Cai, Enhanced photoelectrochemical properties of Cu_2O-loaded short TiO_2 nanotube array electrode prepared by sonoelectrochemical deposition, Nano-Micro Lett. 2 (2010) 277-284. https://doi.org/10.3786/nml.v2i4.p277-284.

[62] D. Lofficial, A. Fecant, D. Uzio, E. Puzenat, C. Geantet, Synthesis and characterisation of $Cu_2O/Pt/TiO_2$ hybrid materials for photocatalytic valorization of CO_2, Photocatalysis for energy [PHOTO4E], 2014.

[63] C. Chen, W. Cai, M. Long, B. Zhou, Y. Wu, D. Wu, Y. Feng, Synthesis of visible-light responsive graphene oxide/TiO_2 composites with p/n heterojunction, ACS Nano 4 (2010) 6425-6432. https://doi.org/10.1021/nn102130m

[64] H. Tada, Q. Jin, H. Nishijima, H. Yamamoto, M. Fujishima, S.i. Okuoka, T. Hattori, Y. Sumida, H. Kobayashi, Titanium (IV) dioxide surface-modified with iron oxide as a visible light photocatalyst, Angew. Chem. Int. Ed. 50 (2011) 3501-3505. https://doi.org/10.1002/anie.201007869.

[65] W. Wu, X. Xiao, S. Zhang, F. Ren, C. Jiang, Facile method to synthesize magnetic iron oxides/TiO_2 hybrid nanoparticles and their photodegradation application of methylene blue, Nanoscale Res. Lett. 6 (2011) 533. https://doi.org/10.1186/1556-276X-6-533.

[66] S. Chaguetmi, N. Sobti, P. Decorse, L. Mouton, S. Nowak, F. Mammeri, S. Achour, S. Ammar, Visible-light photocatalytic performances of TiO_2 nanobelts decorated with iron oxide nanocrystals, RSC Adv. 6 (2016) 114843-114851. https://doi.org/10.1039/C6RA24415G.

[67] J. Liu, J. Ke, Y. Li, B. Liu, L. Wang, H. Xiao, S. Wang, Co_3O_4 quantum dots/TiO_2 nanobelt hybrids for highly efficient photocatalytic overall water splitting, Appl. Catal. B: Environ. 236 (2018) 396-403. https://doi.org/10.1016/j.apcatb.2018.05.042.

[68] K. Arifin, H.A. Kadir, L.J. Minggu, W.R.W. Daud, M.B. Kassim, TiO_2 doped with Fe_2O_3 for photoelectrochemical water splitting electrode: experimental and density functional theory study, Malaysian J. Anal. Sci. 20 (2016) 892-900. https://doi.org/10.17576/mjas-2016-2004-25.

[69] M.A. Mansoor, M. Mazhar, V. McKee, Z. Arifin, Mn_2O_3–TiO_2 semiconducting composite thin films for photo-electrochemical water splitting, Polyhedron 75 (2014) 135-140. https://doi.org/10.1016/j.poly.2014.03.018.

Photoelectrochemical Water Splitting: Materials and Applications
Materials Research Foundations 70 (2020) 129-182

Materials Research Forum LLC
https://doi.org/10.21741/9781644900734-6

Chapter 6

BiVO$_4$ Photoanodes for Photoelectrochemical Water Splitting

Hyungtak Seo[1,2], Basanth S. Kalanoor[3], Shankara S. Kalanur[1]*

[1]Department of Materials Science and Engineering, Ajou University, Suwon 443-739, Republic of Korea

[2] Department of Energy Systems Research, Ajou University, Suwon 443-739, Republic of Korea

[3]Data Scientist Materials Zone, Tel Aviv, Israel

*shankarask@ajou.ac.kr, skshankara@gmail.com

Abstract

Bismuth vanadate (BiVO$_4$) has been significantly utilized as a photoanode component in photoelectrochemical (PEC) water splitting system owing to its excellent optical, electrical properties and suitable band edge positions. BiVO$_4$ has reached a remarkable milestone by exhibiting enhanced PEC activity and O$_2$/H$_2$ evolution rates. In this chapter, the factors such as crystal structure, facet, pH of the electrolyte, carrier dynamics, intrinsic defects, oxygen vacancies, doping, polarons, plasmon-induced resonant energy transfer and surface passivation catalysts influencing the efficiency of PEC water splitting activity of BiVO$_4$ are discussed in detail.

Keywords

BiVO$_4$, Crystal Structure, Facet, Carrier Dynamics, Intrinsic Defects, Oxygen Vacancies, Doping, Polaron, Surface Catalyst

Contents

1.   Introduction

The utilization of metal oxide semiconductor materials in solar water splitting was first successfully reported by Honda and Fujishima in 1972 using n-type TiO_2 [1]. Followed by this report, many oxide semiconductor materials were introduced and extensively researched to explore their suitability for achieving higher efficiency in solar water splitting. However, the subsequent research outcome indicated that binary oxide materials (single junction photoanodes such as TiO_2, WO_3, Fe_2O_3, Cu_2O, etc.) with various limitations were not suitable for water splitting. Consequently, in the process of overcoming the drawbacks of binary oxides, the utilization of ternary metal oxide semiconductors was initiated because they are generally free from the shortcomings faced by the binary metal oxides. In this direction, lot of ternary metal oxides were synthesized, and their water splitting properties were actively investigated. Among the ternary metal oxides, $BiVO_4$ has been the research foci due to its excellent optical and electrical properties feasible for stable and efficient water splitting.

$BiVO_4$ is one of the promising and widely researched n-type semiconductor materials in solar water splitting owing to its excellent optical properties and appropriate band edge position [2]. $BiVO_4$ possesses an indirect band gap of 2.4–2.5 eV capable of absorbing incident visible light up to ~520 nm (including the incident ultraviolet light) [3]. The conduction band edge of $BiVO_4$ is situated near to the value of 0 V potential vs RHE, whereas its valence band edge is positioned at a value of 2.4 V vs RHE [3,4]. That is, the valence band edge position of $BiVO_4$ is located at a significantly higher potential than the solar oxygen evolution potential, and hence during the irradiation, the photogenerated holes provide a sufficiently excess potential for the photo-oxidation of water to oxygen. Simultaneously, the photogenerated electrons under the illumination are capable of maintaining an appropriate potential for the production of hydrogen at the counter electrode with the help of very low applied bias potential (Fig. 1a). Therefore, under ideal working conditions, the $BiVO_4$ photoanode can produce maximum photocurrent of 7.5 mA cm^{-2} when irradiated with standard AM 1.5 G illumination without the help of an added surface catalyst and a corresponding solar-to-hydrogen conversion efficiency of ~9.2% [5]. Importantly, $BiVO_4$ is plentifully abundant, low cost, and exhibits excellent stability during illumination and water splitting conditions. Furthermore, $BiVO_4$ is a nontoxic material and has been estimated to be present in the earth's continental crust of about 135 ppm and 0.17 ppm of vanadium and bismuth, respectively [6,7]. Hence, $BiVO_4$ has been explored extensively as an active photoanode material in PEC water slitting research.

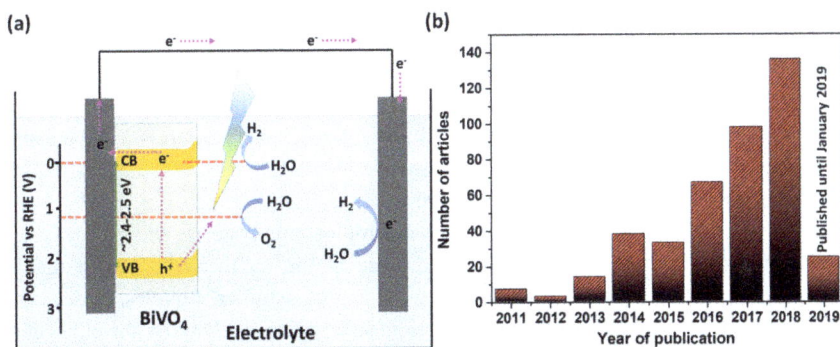

Figure 1: (a) Band edge positions of $BiVO_4$ photoanode. (b) The number of research articles published related to $BiVO_4$ in the last eight years.

Since $BiVO_4$ is an n-type semiconductor material and hence employed as a photoanode in the PEC water splitting system. Because of its n-type character, it exhibits an upward band bending at the interface between electrode and electrolyte. Therefore, under the illumination, the photoexcited electrons are transported towards the counter electrode to carry out the water reduction reaction to generate H_2, whereas the photogenerated holes are utilized in oxidizing water into O_2 at the $BiVO_4$ electrode surface (Fig. 1a). The electrode reaction process undergoing at the $BiVO_4$ surface during the illumination can be defined as shown in reaction 1, whereas the water reduction to H_2 process undergoing at the counter electrode can be represented as shown in reaction 2:

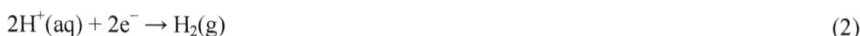

$$2H_2O(l) \rightarrow O_2(g) + 4H^+(aq) + 4e^- \qquad (1)$$

$$2H^+(aq) + 2e^- \rightarrow H_2(g) \qquad (2)$$

The synthesis of $BiVO_4$ was first reported in a patent by Zintl and Vanino for the utilization in therapeutic purposes. Later, $BiVO_4$ was widely used as a "complex inorganic coloured pigments" [8] owing to its brilliant greenish yellow colour. Because of the non-toxic nature and thermal stability, $BiVO_4$ was efficiently utilized as the best alternative to the toxic lead chromate and cadmium sulfide pigments in paint and automobile industries. Further in 1995, Lim et al. have studied and explored the ferroelastic properties of $BiVO_4$ [9]. Importantly, Kudo et al. [2] in 1998 have first demonstrated the successful utilization of $BiVO_4$ as a photoactive material in water splitting to generate O_2 via the water oxidation. In their report [2], efficient production of O_2 was achieved using Ag^+ as a hole scavenger under the visible light irradiation. Since then, tremendous research has been reported and still underway to improve and enhance the solar water splitting efficiently of $BiVO_4$ via implementing various strategies. Fig. 1b illustrates the number of research articles published concerning the solar water splitting activity of $BiVO_4$ (including $BiVO_4$ in heterojunction systems) in recent years. Importantly, Fig. 1b reveals the drastic increase in the number of research articles published recently, indicating that the $BiVO_4$ is gaining significant interest in scientific research groups for its ability to achieve efficient solar water splitting. By the publication trend, it can be estimated that the work on $BiVO_4$ water splitting may witness more significant achievement and milestones in the coming days. Therefore, it is crucial to summarize and highlight the advantageous and critical strategies of $BiVO_4$ studied so far that could assist the future researchers in achieving more significant improvements and milestones.

Interestingly, the PEC water splitting efficiency of $BiVO_4$ suffers from various drawbacks. Specifically, the charge transportation (electrons) in $BiVO_4$ is very slow. Hence, about 60–80% of the electron-hole pairs generated during the illumination were found to recombine before participating in water reduction and oxidation reactions [10]. Also, the weak hole localization, slow kinetics of water oxidation reaction [11] and carrier recombination [12] at the $BiVO_4$ surface were known to affect its water splitting efficiency. Hence, various strategies viz., doping, sensitization with narrow band gap materials, utilization of co-catalysts and the creation of oxygen vacancies etc. have been implemented to overcome the limitations and to enhance its efficiency of water splitting. Importantly, in the last 6 years, enormous progress and milestones have been achieved both experimentally and theoretically regarding the solar water splitting activity of $BiVO_4$. Hence, it has become essential to summarize and highlight the key issues and concepts that influence the solar water splitting activity of $BiVO_4$. Such a study could provide a necessary basis for future research work and ultimately could guide in fabricating much more efficient and stable $BiVO_4$ water splitting systems. Therefore, in this chapter, we focus on the essential strategies established for the design and synthesis of $BiVO_4$ electrode system for the efficient PEC activity.

2.   Crystal and electronic band structure of $BiVO_4$

$BiVO_4$ exists in three distinct polymorphic forms viz., monoclinic, orthorhombic and tetragonal which are mineralogically known as clinobisvanite, pucherite, and dreyerite, respectively [9,13,14]. The crystal structure of $BiVO_4$ exhibits temperature-dependent reversible or irreversible transitions between its phases controlled by one-pair of cation present on the Bi^{3+} [15]. That is, the tetragonal zircon-type $BiVO_4$ was found to undergo an irreversible transformation to the monoclinic scheelite structure between ~350 to 400 °C, and a reversible ferroelastic-paraelastic transition is known to occur at a temperature of 255 °C between the monoclinic-scheelite and the tetragonal scheelite phases [16]. Interestingly, an irreversible transformation from tetragonal to monoclinic phase could be attained via the room temperature mechanical grinding. However, during the mechanical grinding, the amount of conversion was noted to be dependent on the grinding duration [16]. Among the three crystal structures, the monoclinic phase of $BiVO_4$ with superior photocatalytic activity was found to be thermodynamically stable [17]. Hence, the monoclinic phase of $BiVO_4$ has been widely explored and generally employed in PEC water splitting applications.

The monoclinic phase of $BiVO_4$ consists of parallelly stacked Bi–V–O units in a layered structure containing characteristics of both Bi_2O_3 and the V_2O_5 binary equivalents [19]. The base centered monoclinic crystal structure of $BiVO_4$ possesses distinctive lattice locations of Bi (4e), V (4e), O1 (8f), and O_2 (8f) [15]. Hence, the stacking in the monoclinic scheelite structure allows the coordination of Bi atom to O to yield oxygen octahedron with distorted structure, whereas the V is found to be situated in the center of the distorted oxygen tetrahedron [20]. Consequently, in the subunit of monoclinic $BiVO_4$, the distortion caused in the VO_4 tetrahedra creates two different oxygen environments whereas in the dodecahedra of BiO_8, four unique oxygens formed [18]. That is, the O1 is linked to each of V and Bi atoms, in contrast, the O_2 was attached to single V and two Bi atoms [20]. Therefore, the oxidation states of Bi ($5d^{10}6s^2$), V ($3d^0$) and O ($2p^6$) were noted to be 3^+ 5^+ and 2^-, respectively [20]. Importantly, the asymmetric coordination system in $BiVO_4$ was found to be caused by the Bi electronic configuration ($6s^2$) which resulted in sterically active lone pairs [20].

According to the first-principle electronic structure analysis, $BiVO_4$ belongs to direct band gap semiconductors having the ~2.16 eV of band gap (experimental band gap of $BiVO_4$ was described to be ~2.4 to 2.5 eV) with its Fermi level at 2 eV above valence band maximum [18,20]. In $BiVO_4$, the contribution to the valence band comes majorly from the O 2p (comprised of nonbonding O $2p_\pi$ states), coupled to Bi 6s states including a minor contribution from V states. In contrast, the contribution to the conduction band has a major influence from the V 3d states which mainly contains the characteristics of d_{x2-y2} and d_{z2} [18]. The distortions in $BiVO_4$ have been ascribed to the hybridization effects caused by the lone pairs of Bi 6s/O 2p present at the top of valence band. The energy level diagram showing the electronic band structure of $BiVO_4$ is presented in Fig. 2 [18]. Importantly, these couplings in $BiVO_4$ create a notable alteration in band edge structures including downward shift in conduction band edge resulting in shifting of valence band to higher levels. Notably, in the conduction band, the poor overlapping of V 3d orbitals with the Bi 6p orbitals causes the localization of the V 3d orbitals consequently resulting in decreased electron mobility in $BiVO_4$ [7]. Poor electron mobility is known to be one of the crucial factors that affect the water splitting efficiency of $BiVO_4$. Therefore, improving the electron mobility in $BiVO_4$ is essential to enhance its water splitting activity.

$BiVO_4$ is d^{10} material, and hence its full-shell d electrons do not contribute in the bonding process. Mainly, the electrons present in Bi 6s states of $BiVO_4$ forms isolated low-energy bands. A detailed investigation revealed that the existence of the isolated s states and their coupling with O 2p state significantly affect the properties of $BiVO_4$. Importantly, the valence band edge, charge density, and the distortion in the atomic structure have been noticed to be affected. The DFT studies have revealed the narrowing of band gap of $BiVO_4$ and the improvement of its water splitting efficiency through alloying with Bi.

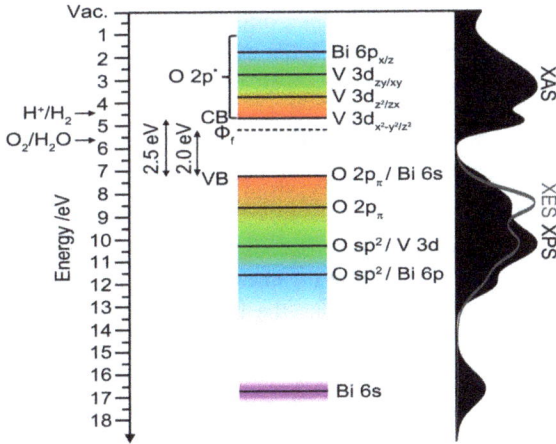

Figure 2: Energy level diagram of monoclinic scheelite BiVO₄ showing the contribution of dominant orbital in valence and conduction bands [18]. In the right side of the image (b) the experimental spectra of X-ray absorption to the empty states of the conduction band and X-ray and photoelectron emission from filled states are presented (Reprinted with permission from American Chemical Society)[18].

3.   The band gap of monoclinic BiVO₄

The clear understanding of BiVO₄ band gap values is highly essential to study its absorption coefficient and excited state lifetime that are crucial for achieving high efficiency in water splitting. However, the determination of BiVO₄ band gap remains a controversy since various reports (computational and experimental studies) debate it has indirect [21–23] band gap, and some suggest it possesses direct [20,24] band gap based on theoretical and experimental studies. Previously, the BiVO₄ was considered as direct bandgap semiconductor based on the density functional calculations [20,25] and later, BiVO₄ was recognized as an indirect band gap through revised theoretical studies [22,25]. However, due to the complicated energetic orbital locations of the Bi 6s states, it is difficult to calculate the band structure using DFT studies [23]. Importantly, the experimental determination of the band gap is highly challenging and, at times, ambiguous as the comprehensive theoretical treatment revealed the presence of direct transitions at slightly higher energy [25]. Recent experimental calculations done by specific spectroscopic techniques have confirmed an indirect band gap of ~2.5 eV (Fig. 3a) for BiVO₄ and at the same time a direct transition was also found to be operative (Fig. 3a) [25]. Therefore, the indirect nature of the BiVO₄ is believed to yield relatively long photocarrier relaxation times, whereas the existence of direct nature confirms the strong optical absorption. Importantly, such a unique optical property of $BiVO_4$ could impose a positive impact on the PEC water splitting activity.

3.1 BiVO₄ photoanode band alignment at a liquid interface

In PEC water splitting, a detailed understanding of the band alignment at semiconductor/electrolyte interface is essential. Notably, the band alignment tends to alter with the change in pH value of the electrolyte. Therefore, the determination of band edge positions of the semiconductor at the interface could help to understand the water oxidation mechanisms undergoing during the PEC water splitting. The pH-dependent band gap and band edges of $BiVO_4$ are represented in Fig. 3b [27]. The band positions of $BiVO_4$ were assigned based on the pH-dependent reaction mechanisms. Conclusively, in the acidic conditions of wide range of pH values the water splitting mechanism was found to be operational. Whereas, in the highly alkaline condition the oxidation of ions takes place which is kinetically favourable. Moreover, under high alkaline conditions, the corrosion could hamper the water splitting activity.

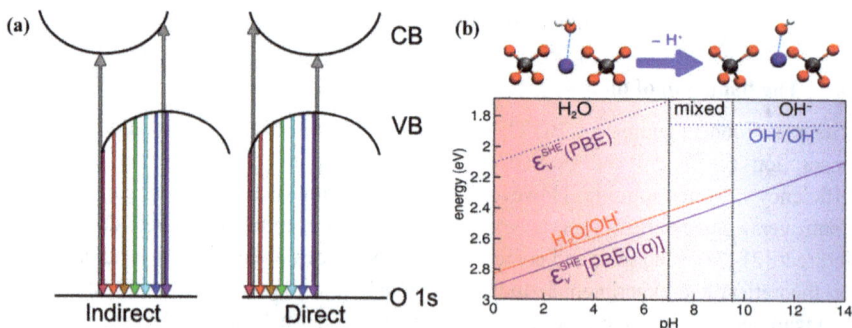

Figure 3: (a) Presentation of the direct and indirect band gap of BiVO₄ [25]. (b) The conduction (ε_c^{SHE}) and valence (ε_v^{SHE}) band edges of BiVO₄ plotted at different pH values. The comparable levels representing water dehydrogenation (red dashed) and the hydroxyl ion oxidation (blue dotted) is also shown [26] (Reprinted with permission from American Chemical Society). The influence of surface coverage from molecularly adsorbed water molecules and hydroxyl ions with respect to the pH is depicted [27] (Reprinted with permission from American Chemical Society).

4. Influence of crystal facet

Generally, certain exposed facets in the metal oxide semiconductor crystals favour photooxidation of water to O_2 whereas some facets prefer photoreduction for H_2 production. In the same way, the PEC water splitting efficiency of $BiVO_4$ is dependent on its surface crystal facets/structure. Because, during the illumination, the photogenerated holes and electrons could be separated, accumulated and undergo catalytic reaction specifically on different facets. Since the different facets in $BiVO_4$ possess specific band energies that produce a driving force for the charge separation and accumulation [28], therefore, facet engineering in $BiVO_4$ crystal structure is highly essential for achieving high water splitting efficiency.

In the $BiVO_4$ crystal, two specific facets were recognized where the water oxidation and reduction reactions occur efficiently. These facets were identified as {010} and {110} via studying the solar water oxidation and reduction reaction via facet-selective photo-deposition of metal and metal oxide catalysts (water oxidation and reduction catalysts) on specific facets (instead of randomly distributed cocatalysts) [28]. During the illumination, the generated electrons and holes get spatially separated between the {010} and {110} facets due to the difference of facet energy levels. Hence, under the irradiation, the exposed {010} and {110} facets on $BiVO_4$ allow efficient charge separation and specifically permit efficient oxidation/reduction reaction separately on the {010} and {110} facets [28]. Using the single-particle photoluminescence spectroscopy and by evaluating the electric-potential-induced photoluminescence properties, the presence of trapped holes was revealed on the {110} facets, whereas the electrons are distributed homogeneously on $BiVO_4$ [29]. Similarly, the $BiVO_4$ photoanode designed with the {040} crystal facet orientation is known to exhibit efficient solar water splitting [30] by providing high photocatalytic active sites. A detailed study [30] revealed that the {040} crystal facet engineered $BiVO_4$ causes the accumulation of photogenerated charges and suppresses the charge recombination that ultimately improves the charge separation efficiency and charge transport efficiency. Importantly, the long-lived charges that are accumulated at the interfacial junction in the geometric structure of the {040} facet favour the water oxidation kinetic processes.

The electrical conductivity of $BiVO_4$ crystal is found to be superior at the ab-plane compared to the c-axis. In terms of crystal facets, the c-axis-oriented $BiVO_4$ indicates the crystal grown along {001} plane and the oriented along b-axis implies growth along {010} direction. The detailed PEC study $BiVO_4$ indicated the high charge transport ability along {010} oriented crystals compared to {001} orientation [31]. The facet oriented physical and electrical properties in $BiVO_4$ are known to be caused by the V ion arrangement in different $BiVO_4$ crystals [33]. Interestingly, the facet oriented electrical conductivity in $BiVO_4$ is dependent on the hopping sites and their influence on the conductivity activation energy. For example, along the c-axis charge carrier transport properties of $BiVO_4$ are governed by hopping mechanism of nearest-neighbour only. In contrast, along the ab-plane, the nearest-neighbour hopping including next-nearest-neighbour hopping is involved. Hence, along the a-b plane, improved conductivity is expected due to the enhanced hopping [33]. Consequently, the $BiVO_4$ photoanodes oriented along {010} direction produce higher photocurrent due to the efficient charge separation compared to {001} orientation (Fig. 4a) [31].

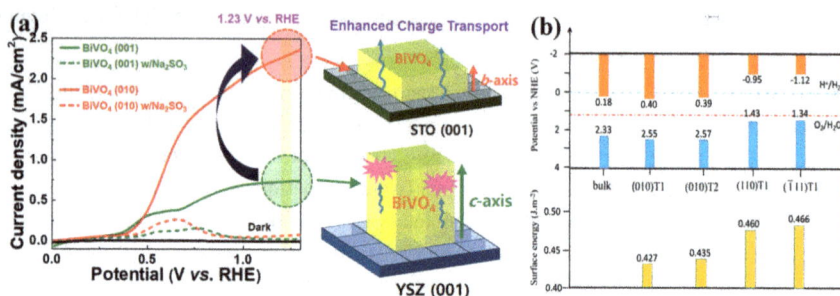

Figure 4: (a) The photocurrent produced by the epitaxial $BiVO_4$ thin films grown in the b-axis and c-axis orientation on STO {001} and YSZ {001} substrate, respectively [31] (Reprinted with permission from American Chemical Society). (b) Facet dependent surface energies and band edge positions of $BiVO_4$ positioned vs normal hydrogen electrode potential at pH = 0 and water redox potentials were also indicated [32] (Reprinted with permission from American Chemical Society).

Interestingly, for the pristine $BiVO_4$ having {001} growth orientation as well as the exposed facets of {001} show superior charge mobility and surface activity [34]. Even though it contradicts the discussion mentioned above, very high photocurrents (~6.1 mA cm^{-2}) were observed in orientation and exposed {001} facet $BiVO_4$ crystals. Compared to the randomly grown $BiVO_4$ photoanode, the photoanode with the {001} growth and facet orientation exhibited higher photocurrent and incident photon to current and charge transfer efficiency [34].

The above-discussed facets of $BiVO_4$ are the low index facets which are more feasible to obtain via synthesis. Theoretically, using the first principle calculation, it is possible to explore the properties of low-index facets including electronic properties, surface reactivity, and stability [32]. In $BiVO_4$ crystal, 4 unique low-index surfaces were reported such as $\{010\}$T1, $\{010\}$T2, $\{110\}$T1, and $\{\bar{1}11\}$T1. These low index facets show different electrostatic potentials which result in dissimilar potentials at the band edge surfaces. Fig. 4b shows the calculated surface energies and edge positions with respected to RHE at zero pH. For reference, the potentials of water oxidation and reduction were also indicated. Hence, the oxidation capability of $\{010\}$T1 and $\{010\}$T2 facets was revealed to be greater than $\{110\}$T1 and $\{\bar{1}11\}$T1. Theoretical calculations confirm the accumulation of photogenerated electrons preferably on $\{010\}$T1 and $\{010\}$T2 faces. In contrast, on $\{110\}$T1 and $\{\bar{1}11\}$T1 facets the accumulation of photogenerated holes is expected. Hence, the BiVO$_4$ crystals having $\{110\}$T1 and $\{\bar{1}11\}$T1 facets were known to be energetically favorable in PEC systems compared to the BiVO$_4$ having $\{010\}$T1 and $\{010\}$T2 facets that contain reduction favorable sites. Because, during the oxygen evolution process on the BiVO$_4$ photoanode, the adsorption of the intermediate is the key step involved as shown below:

$$OH^* \leftrightarrow O^* + H^+ + e^-$$

Importantly, the $\{010\}$T1 and $\{010\}$T2 facets of BiVO$_4$ were found to be thermodynamically stable during solar photooxidation reactions. In contrast, the other surfaces of BiVO$_4$ crystal are reported to be unstable and undergo photo-corrosion under illumination. Also, the $\{010\}$ possesses some interesting properties such as weaker photo-absorption beyond 420 nm, higher mobility of charge carriers, facile water adsorption, and lowers the overpotential for water oxidation [35].

Interestingly water splitting efficiency of BiVO$_4$ not only affected by different crystal facet but also influenced by facet type that is either high index or low index facets. Structure of BiVO$_4$ contains both the low and high index crystal facets. Compared to the low index crystal facets of BiVO$_4$ as discussed above, the high-index crystal facets show enhanced catalytic activity. The BiVO$_4$ having high-index crystal facet provides active catalytic sites [36–38] in comparison with low index facets. For example, during the illumination, a typical 30-faceted BiVO$_4$ polyhedron that is mainly surrounded by high-index facets such as $\{132\}$, $\{321\}$, and $\{121\}$ exhibits 3 to 5-fold enhancement in O$_2$ evolution rate compared to the BiVO$_4$ containing low index facets. However, engineering the BiVO$_4$ having high-index surfaces is difficult, because the growth of high-index surfaces is generally faster than the low index surfaces. As a result, during the crystal growth, the high index surfaces are usually lost due to the minimization of the total surface energy. The theoretical calculation revealed that the dissociation of water is energetically favourable specifically on high index surfaces compared to low index surfaces such as {010}, {110}, and {101} surfaces. Li et al. [39] have synthesized such a $BiVO_4$ polyhedron having high index facets by using the Au ions (formed by the chemical etching of Au nanoparticles) as a structure-directing agent. Thus, formed Au ions were noted to be preferentially adsorbed onto the different $BiVO_4$ surfaces yielding high index facet $BiVO_4$ polyhedron. Using such a high index $BiVO_4$ crystal facets, significantly high O_2 generation quantum efficiency of 18.3 % (under 430 nm light irradiation) has been reported without the use of sacrificial electron donors [39].

5. Carrier dynamics in $BiVO_4$

The dynamics of photogenerated carriers in $BiVO_4$ is known to affect the solar water splitting activity. Mainly, a clear understanding of carrier relaxation as well as trapping is necessary to impose structural modification and improve the water splitting efficiency. Detailed information about the dynamics of photogenerated holes was studied by Ma et al. [40] in the presence of bias potential, electron and hole scavengers through transient absorption spectroscopy. In the space charge layer, the holes having a higher lifetime were known to pose competitiveness between the kinetics of water oxidation and recombination with the electrons in bulk [40]. Fig. 5a shows the kinetic model showing the carrier dynamics of the photogenerated holes participating in either water oxidation or decay. Importantly, two distinct recombination processes were found to be the limiting factor that affects the generation of photocurrent of $BiVO_4$. That is the rapid recombination between the charges (#microseconds) and recombination of holes accumulated at the surface with the electrons in the bulk of $BiVO_4$. Here, the back-electron transfer recombination process takes place in the timescale milliseconds to seconds. Consequently, such recombination in $BiVO_4$ could be significantly avoided by applying a strong anodic bias that provides sufficiently large energetic barrier [40].

The use of broadband transient absorption spectroscopy could allow the measurement of carrier relaxation and trapping rates in the femtosecond to microsecond time scales [41] thus enabling an in-depth understanding of carrier dynamics. During the illumination, the trapping of holes was found to occur in multiple phases. Notably, most of the photo-generated holes in $BiVO_4$ possess trapping time of 5 ps, whereas a significantly minor portion of the holes exhibits trapping time of 120 fs. The photogenerated electrons show the relaxation time of 40s and trapped in 2.5 ns [41]. Furthermore, recombination process which was limited by trapping was found to have a time scale of 10 ns and a spanning time in µs scales. Fig. 5b shows the possible carrier dynamics involved in $BiVO_4$ photoanode. Importantly, spectral and kinetic differences were not observed by applying the bias potential and significant effect of applied voltage occurs only at a longer time scale.

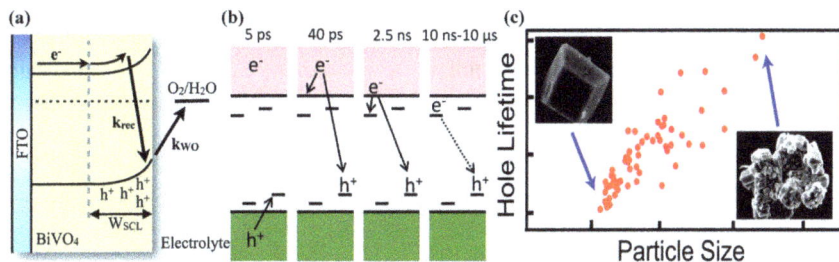

Figure 5: (a) Schematic of carrier dynamics in BiVO₄ photoanode showing the kinetically controlled competition between the back charge recombination with the oxidation of the water process. Where k_{WO} and k_{rec} correspond to the rate constants for water oxidation and back charge recombination, respectively. Here, W_{SCL} represents the space charge region [40] (published by The Royal Society of Chemistry). (b) Model of carrier dynamics in BiVO₄ photoanode showing the lifetime of carriers and their paths [41] (Reprinted with permission from American Chemical Society). (c) The plot of hole lifetime vs particle size showing slower and fast hole decay for aggregated and non-aggregated BiVO₄ crystals, respectively [42] (Reprinted with permission from American Chemical Society).

The photogenerated carrier dynamics and redox reactions of BiVO₄ were also found to be dependent on the geometric structure BiVO₄ crystals. It was revealed that irradiation (527 nm) on well faceted non-aggregated BiVO₄ crystal show comparatively fast hole decay compared to the aggregated BiVO₄ particles having limited grain boundaries [42]. That is, aggregated BiVO₄ consisting of grain boundaries exhibit slower hole decay. Furthermore, the non-aggregated BiVO₄ particles show low reactivity whereas the aggregated particles show significantly high reactivity for the reduction of Fe^{3+}. Fig. 5c depicts the plot of hole lifetime vs particle size revealing slower and fast hole decay for aggregated and non-aggregated BiVO₄ crystals [42]. Interestingly, the hole decay lifetime was noted to have a linear relationship with the secondary particle size of the aggregated BiVO₄ crystals. Importantly, the above discussion concludes that the aggregated BiVO₄ crystals having a significant level of grain boundaries do not function as centers of recombination. In fact, the grain boundaries in BiVO₄ contribute to the increased lifetime of the photogenerated carriers [42]. Therefore, engineering the nanostructure of BiVO₄ crystal is an essential strategy for increasing the carrier lifetime and improving the efficiency of solar water splitting.

6.   Intrinsic defects/Oxygen vacancies in BiVO$_4$

The oxygen evolution process undergoing at the photoanode surface is an energetically uphill reaction that involves a four-electron-transfer process. Hence, the water oxidation is known to be a kinetically sluggish reaction and the bottleneck process in artificial photosynthesis [43]. So, to understand the oxygen evolution kinetics of BiVO$_4$, it is essential to explore its surface properties and the mechanism of bond breaking forming reaction steps that occur on the active sites [7,35,44]. Among the various surface factors affecting the water splitting kinetics, the presence of oxygen vacancies/defect states in BiVO$_4$ is known to influence its water splitting efficiency. The intrinsic defects in BiVO$_4$ can be created in various possibilities and combinations including metal or oxygen enriched sites, antisites, and metal or oxygen vacancies sites [45]. Importantly, the introduction of specific intrinsic defects could allow the tuning of the band gap and to obtain the appropriate BiVO$_4$ structure for the efficient water splitting.

The V and O-rich structures were noted to be favourable for water splitting applications [45]. Here, the V enriched structures refer to the defects created via V-interstitials and antisites on Bi locations, whereas the O-rich structures correspond to the O-interstitials. Interestingly, the Bi-rich, Bi-deficient, and V-deficient structures were noted to be metastable [45]. The Bi-rich structures correspond to the sites containing both Bi-interstitials and Bi-antisites. Fig. 6a represents the various possible defects present in the BiVO$_4$ crystals and their lowest energy configurations (in Fig. 6a the value of x = 0.0625). Importantly, to obtain the BiVO$_4$ with narrow band gap value of 2.3 eV, it is recommended to introduce the defect states with respect to V-antisites on Bi sites along with the Bi vacancies (Bi$_{(1-5x)}$V$_{(1+3x)}$O$_4$, x = 0.0625) [45]. However, along with the band gap narrowing, the defect states presented by Bi$_{(1-5x)}$V$_{(1+3x)}$O$_4$, (x = 0.0625) were found to result in low electron mobility due to the large electron effective masses [45]. In terms of water splitting, the new substitution of V in Bi site could create strongly localized and accommodated d-electrons that result in poor hydrogen evolution reaction on BiVO$_4$ [45].

In addition to the defects created by metal atom alterations, the presence of oxygen vacancies in metal oxide semiconductors could also influence the water splitting efficiency via improving the carrier density/electron transport. In case of BiVO$_4$, the introduction of oxygen vacancies increases the density and mobility of holes allowing the adequate separation of photogenerated electrons and holes [48]. In water splitting, the transport of holes and water oxidation at the electrode/electrolyte interface is the bottleneck process and thus oxygen vacancy introduction in BiVO$_4$ plays a crucial role in controlling the kinetics and catalysis properties. A recent theoretical report on the oxygen vacancies of {040} facet BiVO$_4$ showed some interesting points [46]. For example, the V site on BiVO$_4$ is known to be the active locations for water splitting. Hence, the formation of oxygen vacancies on the $BiVO_4$ surface elevates the active sites of PEC water splitting. Importantly, the presence of oxygen vacancies provides higher adsorption energies of H_2O_{ads}, OH_{ads}, and O_{ads}. Hence, the introduction of oxygen vacancies improves the transfer of holes from $BiVO_4$ electrode into the solution. Prominently, the alteration in Gibbs free energy values confirms that the surface oxygen vacancies of $BiVO_4$ allow the rapid transfer of holes into the electrolyte more favourably compared to the one without the oxygen vacancies. Hence, the oxygen evolution reaction is kinetically and catalytically more feasible on $BiVO_4$ photoanode having significant surface oxygen vacancies [46]. Fig. 6b shows the influence of oxygen vacancies on the water splitting reaction.

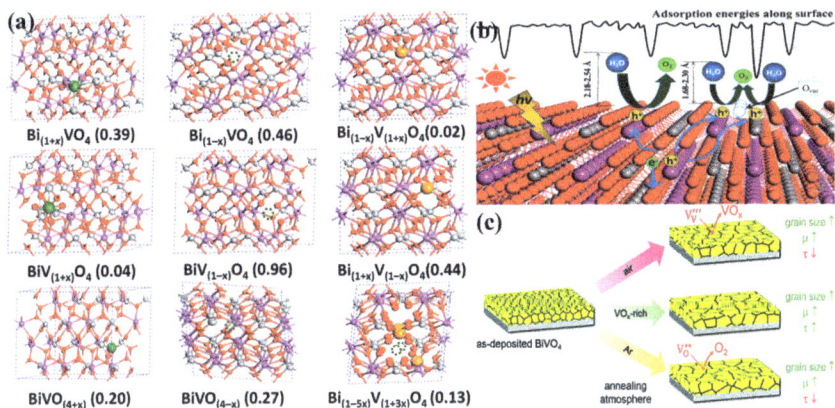

Figure 6: (a) The lowest-energy configurations of different intrinsic defects present in BiVO₄ crystals. The Bi, V, and O atoms are represented in purple, grey, and red colour, respectively. In the BiVO₄ crystal structure, the defects such as interstitial/antisite sites and vacant sites are represented in a green and green circle, respectively [45] (Reprinted with permission from American Chemical Society). The formation energy (in eV) are described next to the molecular formula of the defects below the respective images. (b) The PEC water oxidation process undergoing on the {010} facet of BiVO₄ in presence and absence of oxygen vacancies showing corresponding adsorption energy along the surface [46] (Reprinted with permission from American Chemical Society). (c) Effect of calcination environment on the nanostructure, formation of defects, and charge diffusion in BiVO₄ [47] (published by The Royal Society of Chemistry).

The photocurrent generation by $BiVO_4$ photoanode during the illumination is dependent on factors such as absorption, charge separation, charge transfer efficiency [49]. Notably, the charge transfer efficiency of $BiVO_4$ could be improved by introducing a significant amount of surface oxygen vacancies [50,51]. The abundant oxygen vacancies and the V^{4+} species can be introduced into the $BiVO_4$ photoanode via annealing at high temperatures [52,53] or under H_2 atmosphere [54]. A record photocurrent of about ~5.8 mA cm^{-2} has been reported using duel and multilayers of $BiVO_4$ photoanodes having a significant amount of oxygen vacancies [52,53]. Therefore, having defect states/oxygen vacancies in $BiVO_4$ is highly advantageous. The formation and the suppression of the defects in $BiVO_4$ can be controlled during the heat treatment of $BiVO_4$ photoanodes [47]. Since the carrier transport and recombination in $BiVO_4$ are dependent on the defect states, so by controlling the annealing atmosphere and temperature, the defects can be tuned and consequently carrier transport, and recombination in $BiVO_4$ can be optimized. Fig. 6c shows the formation of defects in $BiVO_4$ photoanode controlled by annealing temperature and atmosphere. Furthermore, the annealing temperature and annealing atmosphere were found to affect the grain size of $BiVO_4$ and its carrier lifetime. For example, the annealing of $BiVO_4$ photoanode higher than 500 ºC causes decreased carrier mobility, and lifetime as a result low photocurrent was recorded. The reduction in PEC properties for the high temperature annealed $BiVO_4$ (500 ºC) was ascribed to the loss of vanadium. Because loss of vanadium at high temperature was found to create deep defect states caused by vanadium vacancy formation. Therefore, the unfavourable creation of oxygen vacancies during annealing could be avoided by carrying out the annealing process in the argon atmosphere that reduces the grain boundaries of $BiVO_4$ by increasing the grain size. As a result, no significant enhancement in the PEC properties was noticed. Similarly, annealing $BiVO_4$ in air limits the disproportionate creation of oxygen vacancies and increases the charge carrier mobility. Hence, air annealed samples show relatively improved PEC properties.

The PEC water splitting efficiency of $BiVO_4$ photoanode having surface oxygen vacancies can be further enhanced by reducing the surface of Bi to metallic Bi. That is, having surface metallic Bi in addition to the surface oxygen vacancies could improve the PEC water splitting properties of $BiVO_4$. The surface reduction can be achieved using $NaBH_4$ solution reduction method [55] or by electrochemical treatment [51]. During the reduction/electrochemical treatment process, both the oxygen vacancies, partially reduced Bi^{3+} and V^{5+} (during electrochemical treatment) [51] and metallic Bi [55] will be created. Here, the presence of both oxygen vacancies and metallic Bi on the surface is highly beneficial. Because oxygen vacancies on $BiVO_4$ will increase the capture ability of photogenerated electrons and improve the adsorption properties of H_2O molecules.

Whereas, the surface metallic Bi/reduced Bi^{3+} effects the photo-response range as well as the separation and consumption of photogenerated carriers, due to the LSPR induced by metallic Bi [55]. Briefly, the presence of both the oxygen vacancies and metallic Bi lead to the extension of visible light absorption range, decreases the charge transfer resistance, and efficiency separates photogenerated charges that ultimately enhances the water splitting efficiency. Similar to the oxygen vacancies, the presence of metal vacancies in $BiVO_4$ was found to influence of PEC properties [56]. Because the introduction of metal vacancy (either Bi or V) changes the current density and valence band edge properties of $BiVO_4$. Specifically, 0.04% of the intrinsic Bi or V defects cause an increase in p-type conductivity [56]. Therefore, the photo-catalytic properties of $BiVO_4$ could also be improved by introducing metallic vacancies. However, a detailed experimental study is required to clearly understand and draw a meaningful conclusion regarding the effect of metallic vacancies in $BiVO_4$.

7.   Polarons in $BiVO_4$

The localization of the excess charge carriers in most of the metal oxide semiconductors causes poor conductivity. Because the localized excess charge carriers create lattice distortions and small polarons. In such situations, the electrical conductively was found to occur via hopping, activated thermally and sensitive to the temperature [57]. Typically, a self-trapped carrier in a potential well polarizes its neighbouring atoms giving rise to polarons [58]. After the generation of polarons in semiconductor lattice, the diffusion of charges takes place from a stable location to another position via the hopping phenomenon. During the course of the migration via hopping, the polarons may get trapped, recombined, or transported to the material surface for catalytic reactions. Consequently, the creation of polarons hinders transport kinetics in the material, and in some cases, it could hamper the photocatalytic process taking place at the interface.

Recent theoretical and experimental reports have suggested that $BiVO_4$ exhibited small polaron [60–62] hopping mechanism causing a decrease in its carrier mobility [63]. A detailed study on the progressive creation and development of polarons indicated that the process of gradual accumulation increased in polarons and the trapping of carriers was complementary [58]. The polaron hopping in $BiVO_4$ was significantly influenced by the thermal distortions [64], and a study of polarons (both electron and hole) produced in $BiVO_4$ at fixed temperatures [59] indicated that the electron polaron was found to localize at V atom. Hence the ionization state of V changes from +5 to +4 state (Fig. 7a). Simultaneously, an increase in the bond length of V−O was observed (Fig. 7a). Similarly, the distribution of hole polaron was found between one bismuth and eight oxygen atoms and causes shortening of Bi−O bond. Importantly, the localization of hole and electrons in the semiconductors causes the creation of polaron states within the band gap [59]. The localized polaron electrons induce one-particle level below the conduction band minimum whereas localized polaron holes occupy a one-particle level above the valence band maximum (Fig. 7b). The polaron levels were found to lie significantly closer at 300k compared to the polaron levels calculated at 0 K (Fig. 7b). Furthermore, the localization of polarons induces significant influence on the band alignment of $BiVO_4$ at the electrode/electrolyte interface and could also cause the recombination of photogenerated charges and thus hinders the photocatalytic oxygen evolution reaction efficiency.

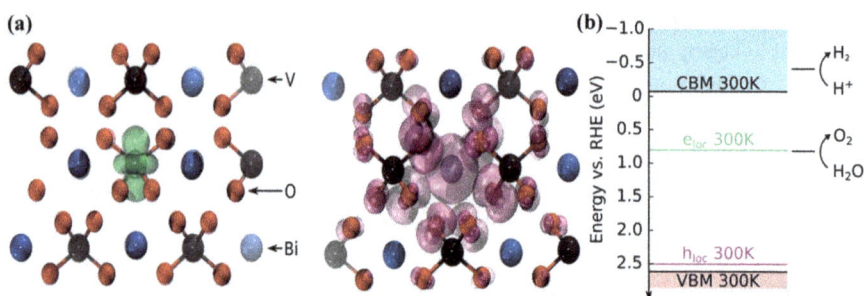

Figure 7: (a) Isodensities of the electron polarons in monoclinic $BiVO_4$ at 0 K (left image) showing the localization of electrons in the vicinity of vanadium [59] and hole polaron (right image) localization around the bismuth atom [59]. (b) band edge alignment of $BiVO_4$ and transition levels of polaron at 300 K plotted with respect to water oxidation and reduction pH = 7 [59] (Reprinted with permission from American Chemical Society).

The hopping barrier (E_a) of the polarons created in $BiVO_4$ was found to be dependent on the lattice constant, operating pressure and temperature conditions. The excess electrons localized on the V atom cause substantial distortions in the vicinity of reduced V^{4+} positions [61]. A study by Kweon et al., [61] suggested a value of hopping barrier of ~0.35 eV for a small polaron. The hopping barrier was found to grow linearly with a lattice constant due to the changes in pressure and temperature. For example, at α = 15%, the small polaron state is predicted to be about 0.5 eV more favourable than the completely delocalized state [61]. Importantly, the polaron hopping barrier E_a was found to increase monotonically from 0.10 eV to 0.46 eV as α increases from 5% to 25%.

Hence, a substantial increase in the electronic energy gain E_e relative to the strain energy loss E_s was observed [61].

It is interesting to assess the $BiVO_4$ properties in case of excess electron delocalization and their relative polaron behaviour. In highly electron-doped $BiVO_4$, the polaron–polaron separation is sufficiently small. Hence the overlap between the diffusing electron wavefunctions was expected that causes concentration dependence of electron mobility. The process of doping metal ions including Mo- and W- was found to introduce additional electrons in $BiVO_4$. Note that, upon doping the creation of polarons was observed in bulk and as well as on the surface of $BiVO_4$. Generally, doping leads to the formation of self-trapped small polarons in $BiVO_4$. In case of $BiVO_4$ doped (by Mo or W) at the surface, the polarons prefer to be localized on the transition metal ions present in the subsurface layer. More precisely, the location of polaron was found to be different in Mo and W doped $BiVO_4$. Upon Mo doping, the electron polaron remained on Mo beneath the surface layer. However, in the case of W doping, the electron polaron was believed to be near V ions. In general, the electron polaron is expected to remain beneath the surface layer. The localized positions of polarons in $BiVO_4$ were rationalized by the energy levels of d-orbitals of transition metals as well as the variation of electrostatic potential [65]. Here, the localized polarons present in the subsurface layer in doped $BiVO_4$ could significantly improve the feasibility of electrons to the significant catalytic sites on the $BiVO_4$. That is, the presence of polarons in the subsurface layer is believed to build a fast lane for electron migration to the surface reaction sites. Furthermore, this lowers the possibility of recombination between photogenerated charges. As a result, notable improvement in photocatalytic properties are expected in electron-rich doped $BiVO_4$ systems [65].

In case of orthorhombic $BiVO_4$, a highly anisotropic small hole polarons state was found at the conduction band edge. Similar to monoclinic structure, the orthorhombic $BiVO_4$ also possesses local lattice distortions in VO_4 tetrahedral symmetry leading to the polarons contributed by the vanadium 5+ and 4+ states. This indicates that, in addition to monoclinic $BiVO_4$, the orthorhombic $BiVO_4$ could also act as a vital visible-light photocatalyst. However, the "charge trap" scenario exhibited by polaronic state expected to be a fundamental issue and the primary cause of poor water splitting performance [66].

8.   Doping $BiVO_4$

Generally, the process of doping in metal oxides induces a significant change in morphology, crystal structure, optical, electrical and band edge properties. Interestingly, optimized doping causes positive changes which assist in improving PEC properties and in some cases could also cause negatives effects. Therefore, it is essential to investigate and optimize the level of doping in metal oxides for the enhancement in solar water splitting properties. The PEC water splitting activity of $BiVO_4$ photoanode faces severe limitation due to its poor bulk charge transport and the presence of polarons as discussed earlier. Therefore, doping of metal ions into $BiVO_4$ effects the inter-band gap polaron states as a result photoconversion efficiency will be affected. The impact and extent of doping on the various properties of $BiVO_4$ including the polarons are dependent on the type of metal ion dopant and its concentration. In this section, the effect of various dopants on the PEC properties of $BiVO_4$ is discussed. Till now, different dopant materials were utilized to improve the photocatalytic properties of $BiVO_4$. Among the dopants, W and Mo have been extensively studied and established as the most suitable dopants for $BiVO_4$. Therefore, it is essential to explore the insights of the doping effects of W and Mo on $BiVO_4$ and corresponding photocatalytic properties.

8.1   W doping into $BiVO_4$

Tungsten (W) ion is one of the widely used dopant material for $BiVO_4$ photoanode for enhancing its PEC water splitting activity. Therefore, it is essential to explore the effect of doping on the electrical properties of $BiVO_4$ photoanode. When doped, W ions were found to occupy V sites [67,68] in $BiVO_4$ crystal (Fig. 7a). This results in the reduction of Bi-ligand distance and causes less distortion in the local structure at the Bi center. Hence, the bulk structure monoclinic sheelite was found to change slightly into a tetragonal structure yielding the mixture of monoclinic and tetragonal structures [67]. Doping of W alters the nature of hole trap states, and hence suppression of carrier recombination, as well as an increase in the lifetime of electrons, was observed. Fig. 7a presents the carrier dynamics of W doped $BiVO_4$ compared with the undoped WO_3 showing the removal of the majority of intrinsic hole traps. As a result, $BiVO_4$ photoanodes doped with W exhibit higher photocurrent compared to undoped ones. Doping of W does not significantly improve the carrier transport of $BiVO_4$, whereas the increase in photocatalytic properties solely ascribed to the improved charge separation and carrier lifetime [67].

A recent experimental and theoretical analysis revealed that doping of W does not affect the morphology, crystallinity and optical properties of $BiVO_4$ [69]. So, the significant enhancement in the photocurrent was attributed to the increased carrier density, electron/hole separation and active transport of charges towards the material surface [69]. Conclusively, the electrochemical studies have confirmed the decreased charge transfer resistance at the $BiVO_4$/electrolyte interface and the production active surface states for charge transfer upon W substitution [69]. Theoretical studies indicated that the doping of W helped the activation of the V enabling them to act as active reaction sites. Hence, upon W doping, the surface-active water oxidation location changes from the Bi atom to the V atom. Furthermore, W doping decreases the adsorption energies, bond lengths between the surface atoms, and distance of active adsorption species such as O_{ads}, OH_{ads}, and OOH_{ads} [69]. The stoichiometry of W doped $BiVO_4$ system is one of the curial factors for achieving the high photocatalytic activity. In stoichiometric $BiVO_4$, the charges of Bi, V and W atoms were found to be balanced while in non-stoichiometric $BiVO_4$ the excess of either Bi or Bi and W. Interestingly, the band gap of $BiVO_4$ did not show any significant change either in the stoichiometric or non-stoichiometric ($BiV_{1-x}W_{x+\delta}O_4$) compositions as previously discussed. However, non-stoichiometrically doped $BiVO_4$ exhibited higher photocurrent compared to the stoichiometrically doped $BiVO_4$. In non-stoichiometric doping, that is in excess of either Bi or Bi and W, W^{6+} was found to replace the V^{5+} without generating other defects. Furthermore, the excess of Bi and W in the $BiVO_4$ system decreased the trap locations and improved the charge transfer kinetics [70].

8.2   Mo doping into $BiVO_4$

Like W ion, the doped Mo atom also situates at the $BiVO_4$ crystal surface via substituting for V atom in the tetrahedral sites [72]. The substitutional doping of Mo causes a reduction in vanadium (V^{5+} to V^{4+}) and produces intraband gap states. The doping of Mo also causes the deformation crystal due to the bigger VO_4 tetrahedral. The binding energy value of the introduced occupied state was found to be 1.4 eV, positioned above the valence band maximum at 0.9 eV. This intraband state is related to the creation of small polaron. Doping of Mo into $BiVO_4$ brings additional electrons which induce further distortion in VO_4 tetrahedral. Such deformations produce the sites having either V d_z^2 splitting or the V d_{zx} and V d_z^2 hybridization. Hence new conduction band states were created upon Mo doping. By theoretical and spectroscopic data, it was revealed that the Mo doping induced polaron was situated between the valence band maximum and the oxygen evolution potential. Importantly, the formation and location of the polaron state produced a strong influence on the charge mobility and recombination at the surface that in turn effected PEC water splitting activity [72].

Subsequent research on Mo-doped $BiVO_4$ indicated that Mo doping enhanced the PEC properties of $BiVO_4$ [71]. For example, a Mo-doped $BiVO_4$ photoanode having inverse opal structured (Fig. 8b) showed increased PEC water splitting efficiency compared to undoped $BiVO_4$ [71]. Experimentally, the kinetic efficiency of charge separation and oxidation were found to be much higher in Mo-doped $BiVO_4$ compared to undoped $BiVO_4$ [73]. Furthermore, Mo-doped $BiVO_4$ generated higher photovoltage at the electrode/electrolyte interface [73]. Note that, the extent of enhancement in PEC properties of $BiVO_4$ upon Mo doping was dependent on the amount of Mo doping. About 2% Mo doping exhibited best photocatalytic performance [74], and the conduction band edge showed an upward shift towards more negative potentials. Because, 2% Mo-doped $BiVO_4$ showed highest donor concentration (5.7×10^{27}) and decreased flat band potential of -0.45 V, compared to the other doped amounts. Importantly, the increase in the donor concentration resulted in the shift of the Fermi level upwards which allowed efficient electrons transport in Mo-doped $BiVO_4$ compared to undoped $BiVO_4$ [74].

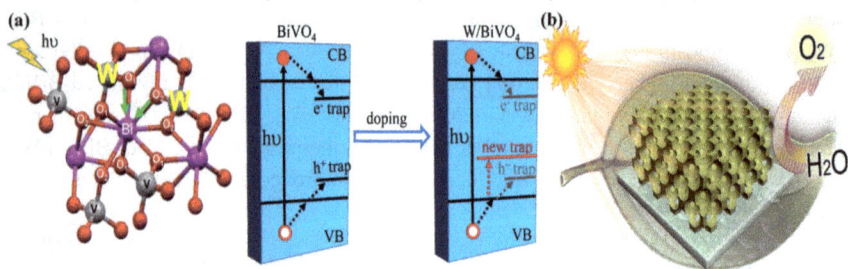

Figure 8: (a) Crystal structure of $BiVO_4$ showing the W atoms occupying V sites. Carrier dynamics of undoped and W-doped $BiVO_4$ revealing the change in hole trap states [67] (Reprinted with permission from American Chemical Society). (b) Inverse opal structured Mo-doped $BiVO_4$ photoelectrode as an artificial leaf for photoelectrochemical water splitting [71] (Reprinted with permission from American Chemical Society).

The photocurrent of Mo-doped $BiVO_4$ can be further enhanced by tuning its structure and morphology [75]. Using the Gartner model, Rajini et al. [76] showed the increased hole diffusion length to 183 nm from 47 nm for Mo-doped $BiVO_4$ compared to from undoped $BiVO_4$. Also, doping of Mo into $BiVO_4$ resulted in better charge separation as confirmed by open circuit photovoltage decay measurements [76] due to the increased hole diffusion length which suppresses the recombination of photogenerated charges. Further, a significant decrease in the charge transfer resistance was noted [76]. In case of tetragonal $BiVO_4$, the high amount of Mo doping was found to introduce cation vacancies (Bi^{5+})/oxygen interstitials. Here, the presence of Bi^{5+} states is believed to act as a catalytic water oxidation site in Mo-doped $BiVO_4$. As in the literature, the standard redox potential of Bi^{3+}/Bi^{5+} and water oxidation potential O_2/H_2O were located at a potential of $+1.59$ V and $+1.23$ V at pH 0, respectively [14,77]. That is, the redox potential of Bi was situated at more positive potential than the water oxidation potential. Therefore, the presence of Bi^{5+} state tended to regulate the redox potential values according to water oxidation potential [78]. Such a redox potential alignment was known to increase the water oxidation rate and hence improved photocatalytic activity was expected in Mo-doped $BiVO_4$ photoanodes [71].

8.3   Other dopants in $BiVO_4$

In addition to W and Mo, many other materials such as Nb [79], Y [80], Co [81], C [82], In [83], Se [84], P [85], B [86], N_2 [48], and F [87,88] have been utilized as dopants in $BiVO_4$ photoanode. Briefly, 10% of Nb doping in $BiVO_4$ drastically enhances the photocurrent, H_2 generation and dye degradation properties without showing any significant change in optical properties (band gap) have been observed [79]. The doping Y into $BiVO_4$ at moderate level was noticed to change the monoclinic structure into a mixture of the monoclinic and tetragonal phases. However, after the doping of high amounts of Y in $BiVO_4$ was found to convert the crystal phase completely into the tetragonal structure. As a result, doping of Y directed the conduction band to move in the negative direction and hence the increase in the photocurrent and dye degradation ability was observed [80]. The charge separation in bulk and at the interface can also be improved by providing a Co-doped $BiVO_4$ layer on the Mo-doped and undoped $BiVO_4$ which showed high photocurrent and water splitting activity [81]. Doping of optimized amount (1.5 wt%) of C in $BiVO_4$ has also been reported to produce enhanced O_2 evolution and dye degradation capability [82]. Similarly, doping of In was reported to increase the photocurrent of $BiVO_4$ photoanodes. However, unlike W and Mo, the doping of In takes the Bi^{3+} site in the $BiVO_4$ structure instead of V sites. Doping of In did not cause any change in the crystal phase, band gap, and morphology [83]. The band gap of $BiVO_4$ can be significantly reduced by doping Se ion. In fact, the doping of Se could also induce an upward shift in both the conduction as well as valence band positions. Therefore, the doping of 1–2% of Se could be an effective strategy for tuning the band gap and band edge properties of $BiVO_4$ [84]. As previously discussed, the introduction of oxygen vacancies in $BiVO_4$ has been an effective strategy for improving its water splitting activity. Nitrogen-doped $BiVO_4$ can be obtained by treating $BiVO_4$ in the N_2 flow at high temperatures (350 ºC) [48]. During the process of nitrogen doping, the oxygen vacancies were also introduced. The doping of nitrogen reduces the band gap $BiVO_4$ and increases the carrier density and mobility. Hence, improved charge separation and photocurrents were obtained. The abundant amount of oxygen vacancies in $BiVO_4$ can also be induced via P doping. An optimized amount of P, that is 2% of P doped $BiVO_4$ showed high carrier concentration and decreased charge transfer resistance by increasing the number of oxygen vacancies. Consequently, doping of P in $BiVO_4$ improved its water splitting performance [85]. Like W dopant, substitutional doping of B (0.6%) [86] improved the PEC properties of $BiVO_4$ without altering its band gap energy.

Unlike W and Mo, the insertion of Sn into $BiVO_4$ is found to occupy Bi sites instead of V sites [68]. Because, the properties such as distribution of electrons in the outer shell and the ionic radius of Sn show significant mismatch with the host atoms and thus causes an increase in the formation energy compared to dopants such as Mo or W and causes, segregation of SnO_2 on Sn^{4+} surface [68]. Therefore, Sn-doped $BiVO_4$ exhibits weak PEC water splitting activity compared to the W or Mo-doped $BiVO_4$ [68]. Therefore, Sn doping into $BiVO_4$ is not suitable for PEC water splitting applications. Furthermore, we can conclude that the metal ion doping that involves the replacement in the Bi site is not favourable in water splitting. Importantly, the dopant metal ion for $BiVO_4$ should possess a comparatively high valence state, low formation energy and should not form segregation on the surface to yield better water splitting efficiency. The PEC activity of $BiVO_4$ can also be improved by doping with F. The incorporation of F inside $BiVO_4$ was found to substitute interstitial oxygen positions connected with vanadium and bismuth defects [88]. The doping of F forms stable structure and introduces cation vacancies [87] in $BiVO_4$ without any phase transition [88]. Similar to the Se doping, the band gap value of $BiVO_4$ showed slight decrease upon F doping [88]. Therefore, the incorporation of F improved the PEC performance of $BiVO_4$ [87].

8.4   Lanthanide ion doping into $BiVO_4$

Unlike most of the metal dopants, the lanthanide elements show different characteristics upon doping to $BiVO_4$. Mainly, the substitutional doping of lanthanides such as La, Ce, Sm, and Yb takes place at the Bi^{3+} site instead of V site. Therefore, the alteration in the electrical properties of $BiVO_4$ caused by doping was noted to be different for different lanthanides. All the lanthanide doped $BiVO_4$ photoanodes did not show an increase in the photocurrents. Mainly, the doping of Sm and Yb increased the photocurrent whereas doping of La and Ce showed decreased photocurrent compared to the undoped $BiVO_4$. The increase in the effective mass of charges (electrons and holes) causes a decrease in the carrier mobility of $BiVO_4$ upon La doping thus show decreased photocurrent. In case of Ce doped $BiVO_4$, the reduction in photocurrent values was also ascribed to the increased effective masses of charge carriers. Additionally, Ce doping introduces filled midgap region near the valence band maximum of $BiVO_4$. Such midgap states act as recombination centers and enhance the recombination of photogenerated charges. In contrast, the doping of Sm and Yb increases the carrier density in $BiVO_4$. Therefore, oxygen vacancies were introduced into the system giving rise to the donor states without any significant shifts in the valence and conduction band edges. These donor states were found to be located below the conduction band minimum. Therefore, the enhanced photocurrents were interpreted as a synergic effect due to the doping of Sm/Yb and oxygen vacancy formation [89]. However, another report by Javier et al. [90] showed that doping of La (1%) and Ce (2%) enhanced the photocurrent of $BiVO_4$. As this contradicted the previous discussion; a detailed investigation is required to clearly understand the effect of doping with respect to dopant concentrations.

8.5   Codoping in $BiVO_4$ (multiple ion doping)

As discussed in the earlier sections, doped $BiVO_4$ photoanode with a single type ion showed significantly improved PEC properties compared to the undoped $BiVO_4$. Interestingly, doping $BiVO_4$ simultaneously with two different ions could further enhance its PEC properties. For example, doping Ti along with W has shown enhanced water splitting activity compared to only W doped $BiVO_4$. The co-doping of W and Ti causes increased charge separation, hole transport and leads to a negative shift in the onset potential. Importantly, the activation energy required for the hole polaron hopping inside $BiVO_4$ could be decreased (11.5%) significantly by doping with both W and Ti [91]. That is, the required activation energy was revealed to be lower than in case of W doped $BiVO_4$ [91]. Furthermore, a significant increase in the hole mobility of about 29% was achieved upon codoping with W and Ti [91]. The Ti sites on the $BiVO_4$ surface were noted to be energetically more favourable for water splitting compared to W sites [91]. Compared to undoped and W-doped $BiVO_4$ photoanode, the W-Ti doped $BiVO_4$ possesses lower overpotential required for water splitting. Similarly, co-doping using both W and Mo resulted in higher PEC activity compared to the individually doped $BiVO_4$ photoanode [92]. It can be concluded that co-doping with multiple ions could be much more effective strategy to enhance the PEC properties of $BiVO_4$ compared to the single ion doping [91].

9.   The side of illumination on $BiVO_4$ photoanode

During the PEC water splitting, the side of incident light illumination on the photoelectrode is very crucial. For the PEC water splitting system the term front illumination and back illumination were used when the incident radiations were passed from the photoanode side and FTO side, respectively [93–95] (note that, in some reports, the reverse terminology has beenused in assigning the side of illumination). In case of $BiVO_4$ photoanode, the photocurrent produced during the illumination depends on the factors such as electron diffusion and thickness of the film and side of irradiation. With the mesoporous $BiVO_4$ thin films of different thickness, it was revealed that front side illuminated photoanodes showed higher photocurrent compared to backside illuminated samples. The significant suppression of charge recombination was recorded due to the efficient charge separation during the front side illumination compared to the back side illuminated samples [96]. During irradiation, the photogenerated electrons travelled from

$BiVO_4$ film towards the FTO substrate. Hence, during the back-illumination scenario, a trap free electron transport region was found to exist in the vicinity of the FTO substrate/$BiVO_4$ interface. At the same time, a trap limited transport region was found to exist between the trap free electron transport region and the electrolyte interface (that is away from FTO substrate). Importantly, the effective diffusion of photogenerated electron towards FTO occurs from the nearest $BiVO_4$ due to the relatively short pathways in the mesoporous film.

In contrast, during the back illumination, only a trap limited transport existed throughout the $BiVO_4$ film [96]. Because the electrons generated in the vicinity of electrolyte during the irradiation have to travel across the number of unfilled trap states. Fig. 9a shows the regions of the electron free transport and trap limited transport present in the $BiVO_4$ film with respect to the illumination side [96]. Therefore, higher photocurrents were expected to produce during the light exposure along FTO side, whereas the illumination via photoanode side showed comparatively lower photocurrent values. The simulation studies revealed that near the FTO surface region, higher electron density was detected which were tolerant towards the recombination. Such an electron density or the existence of the recombination tolerant region was not possible during photoanode side illumination. However, after considering the thickness factor, FTO side illumination has been recommended for the thicker $BiVO_4$ films whereas photoanode side illumination has been favourable for ultrathin films because of the interior light intensity distribution.

10.   Photo-charged BiVO₄

Trześniewski and Smith [97] have proposed a unique way to enhance the water splitting activity of $BiVO_4$ photoanode known as photo-charging [98]. Remarkably, these photo-charged $BiVO_4$ photoanodes have shown enhanced photocurrent, photovoltage and cathodic shift of onset potential compared to untreated electrodes [97,98]. The photo-charged $BiVO_4$ photoanodes could be obtained by exposing the electrodes to AM 1.5 G illumination for long intervals under open circuit configuration. The treatment of photo-charging was found to suppress the surface recombination and improves both the bulk and surface properties by passivating the surface defect states of $BiVO_4$. Fig. 9b shows the change in band alignment and band bending before and after the introduction of surface defects in the presence and absence of illumination on the metal oxide semiconductors in general. When photoelectrode comes in contact with the electrolyte (under dark), the Fermi level (E_F) of the photoanode aligns with respect to the potential level of water oxidation and equilibrates. In the absence of surface defects, the light illumination on the electrode tends to shift the quasi-Fermi level of the electrons ($E_{F,n}$) near to the conduction band edge (E^S_C). At the same time, the quasi-Fermi level ($E_{F,p}$) of holes aligns with the redox potential of $E^0_{OH^-/O2}$ (Fig. 9b). However, in the presence of surface defects, the band alignment and bending behave differently under the dark and illumination conditions. The presence of sufficient density of surface states (10^{13} to 10^{14}) [99] at the interface between the photoelectrode and solution, causes a considerable drop in potential at Helmholtz region [100] under dark. Hence, the Fermi level aligns with the energy level of the electronic surface states (E_{SuS}) instead of the potential level of $E^0_{OH^-/O2}$ [101,102]. Here, the potential drop observed at the Helmholtz layer is known as the Fermi level pinning effect [101,103]. Therefore, under illumination, both Fermi level and degree band bending do not show significant change resulting in lower photovoltage values. The above discussion concludes that the process of photo-charging passivates the surface states and the subsequent effect of Fermi level pinning reduces and photovoltage increases. It was found that the photo-charging will lead to the reduction of V from 5+ state to 4+ in the bulk of the film [97,98].

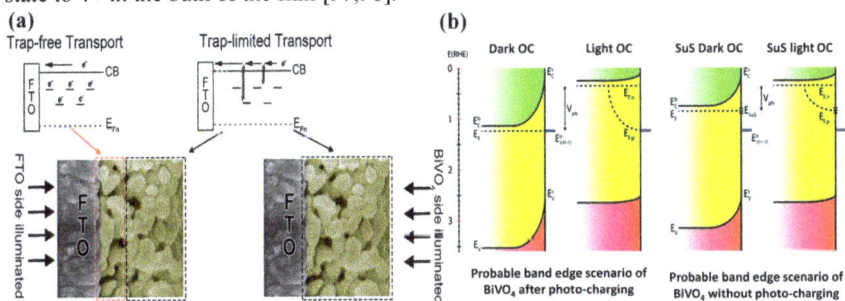

Figure 9: (a) The presence of trap free electron transport region in the vicinity of FTO and the trap limited transport region away from FTO substrate during the illumination from FTO side [96] (Reprinted with permission from American Chemical Society). Similarly, the presence of only trap limited region present in the BiVO₄ film during the illumination from the photoanode side [96]. (b) The difference in band edge alignment of BiVO₄ under dark and light with and without photo-charging [97]. Here the terms OC refers to open circuit configuration, and SuS refers to electron surface states [97] (Published by The Royal Society of Chemistry).

11. Hole blocking layer for BiVO₄

The insertion of a metal oxide layer between the BiVO₄ and the transparent conductive oxide (generally ITO or FTO substrate) in a photoanode is one of the important strategies utilized for improving the PEC properties. Such an inserted layer is termed as a hole blocking layer or underlayers for BiVO₄. Not much work has been reported on such underlayers or hole blocking layers particularly for $BiVO_4$. Recently, metal oxides such as SnO_2 [104–106] and Lu_2O_3 [107] have been utilized as underlayer/hole blocking layers for $BiVO_4$ photoanodes. The underlayers for $BiVO_4$ allowed the efficient flow of photogenerated charges yielding efficient charge separation due to the favourable band bending at the electrode/electrolyte interface. Furthermore, the metal oxide layers could also act as a passivation coating [108–111] for the trap states and the interface causing less recombination of the photogenerated charges. [108,109]. Therefore, it is highly beneficial to install a hole blocking layer/underlayer between $BiVO_4$ and conducting substrate.

A recent study by Segi et al.., [104] has revealed that the photocurrent of $BiVO_4$ photoanode was drastically increased by inserting a hole blocking layer of SnO_2 [108,109]. Interestingly, the thickness of SnO_2 underlayer tends to affect the intrinsic property of $BiVO_4$ layer. That is, the alteration in the SnO_2 thickness causes a significant change in the stoichiometry between Bi/V, band gap, crystal structure, grain size, and film roughness. The $BiVO_4$ film deposited on FTO substrate in the absence of SnO_2 layer or presence of thin (20 nm) SnO_2 layer appears rough consisting of large grains. In case of thicker SnO_2 layer, the nucleation of $BiVO_4$ during deposition happens on larger SnO_2 site instead of FTO sites. Hence, the $BiVO_4$ formed on the thicker SnO_2 layer showed smaller crystalline sizes [104]. In the absence of SnO_2 underlayer, the photogenerated charges are expected to be recombined due to the presence of trapping states at the interface of FTO/$BiVO_4$ [104]. Hence, the $BiVO_4$ photoanodes show poor PEC properties in the absence of SnO_2 underlayer. These defect trapping states are believed to be exposed even in case of a thin layer of SnO_2, thus causing recombination of charges and decreasing the water splitting efficiency. However, the presence of thicker SnO_2 underlayer passivates the defect trapping states present at the interface and thus suppresses the recombination and yields higher efficiency during water splitting. As discussed above, the interpretation of PEC properties with respect thin SnO_2 layer is not always the same. Because, by using a very thin layer (8 nm), containing low defect density is found to enhance the PEC properties [106]. However, such a high-quality, low defect density SnO_2 could be obtained via an atomic layer deposition technique [106].

The PEC properties of $BiVO_4$ photoanode consisting of SnO_2 underlayer can be further enhanced by slightly tuning the stoichiometry of Bi and V together with Mo doping [105]. By simultaneously optimizing Bi and V ratio and Mo content in $BiVO_4$, a favourable band alignment between SnO_2 and $BiVO_4$ can be obtained. The configuration having a small conduction offset was found to exhibit improved PEC properties. A recent report suggests that an ultrathin film of Lu_2O_3 can also be used as an underlayer/hole blocking layer between the interface of transparent conducting substrates and $BiVO_4$. By an atomic-scale interface control of Lu_2O_3 an efficient transport of carrier can be achieved. Importantly, the unique band alignment of Lu_2O_3 underlayer was found to block the photo-generated holes [107]. Hence, efficient separation of photo-generated charges was attained which enhanced the PEC properties. Importantly, the underlayer of Lu_2O_3 interlayer modulates the electronic conduction pathways along the structural grain boundaries [107].

12. Catalyst coatings on BiVO₄ photoanode

The interface of photoanode and electrolyte controls the kinetics of charge transfer, and hence the presence of a defect and strap states at the photoanode surface participate in high surface recombination process that significantly improves water splitting efficiency [101,112,113]. Generally, $BiVO_4$ photoanode is coated with a variety of materials which show multiple activities such as protection to the photo-corrosion, electrocatalyst, light trapping layer, hole extraction layer and suppression of recombination by passivating defect states. Because the photoanodes do not possess enough catalytic activity sites for the water oxidation reaction on the surface, therefore, it is essential to passivate the photoanode with suitable catalyst material. Hence, the coated $BiVO_4$ showed enhanced photocurrent, IPCE and H_2/O_2 gas production rates compared to uncoated $BiVO_4$. In fact, a major portion of the recent research has been directed for exploring various coating materials on $BiVO_4$ to enhance its PEC properties. Importantly, very high photocurrents were achieved reaching an important milestone using $BiVO_4$ photoanodes in water splitting. In this section, various coating layers employed on $BiVO_4$ and their PEC results are summarized. Table 1 lists a variety of coating materials utilized on the $BiVO_4$ and their respective, photocurrent, IPCE and H_2/O_2 production values.

The carrier dynamics taking place at the electrode interface in the absence of a catalyst as well as in the presence of surface passivation catalyst and the non-passivating catalyst is shown in Fig. 10. In the absence of catalyst on $BiVO_4$ photoanode, the holes generated under illumination in the valence band will either transport to electrolyte interface to carry out water to O_2 reaction or recombine with conduction band electrons at the surface states as shown in Fig. 10a [164]. Since the water oxidation catalysis of $BiVO_4$ surface is poor, the photogenerated holes are more favourable to recombine, and thus water splitting efficiency decreases significantly. By using the Co-Pi surface passivation catalyst (Fig. 10b) on $BiVO_4$, the surface recombination of photogenerated holes at surface states can be suppressed. Hence, the photogenerated holes are injected into the electrolyte to oxidize water to O_2 directly. However, due to the porous nature of deposited Co-Pi catalyst, a portion of $BiVO_4$ will be exposed directly to electrolyte [164], and the poor kinetics of water oxidation reaction of $BiVO_4$ surface will limit the direct ejection of photogenerated holes into the electrolyte. As a result, the photogenerated holes will take an alternate but competing pathway via Co-Pi catalysts as shown in Fig. 10b [164]. Importantly, the charge transfer rate constant decreases with applied potential due to the low thermodynamic driving force. The possible carrier dynamics happening in the case of non-passivating RuO_x co-catalyst on $BiVO_4$ photoanode is shown in Fig. 10c [164]. Here, the presence of RuO_x does not affect charge recombination of photo-generated holes due to the absence of passivation, and thus no significant enhancement is expected in water splitting activity [164]. Therefore, the choice of a suitable passivating and catalyst layer on $BiVO_4$ is highly crucial for achieving high PEC water splitting efficiency [164].

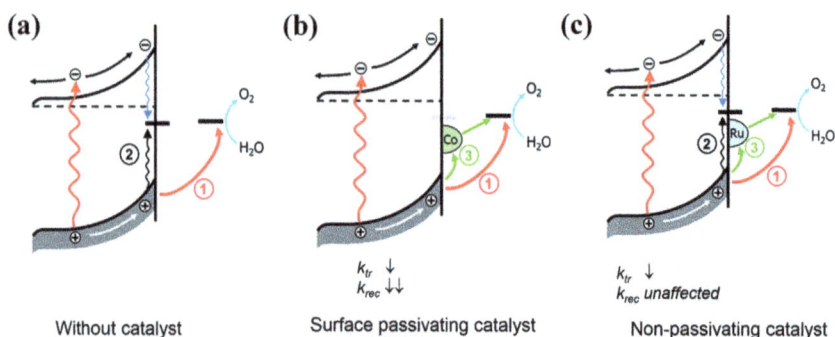

Figure 10: Model of carrier dynamic process involved in the BiVO₄ photoanode/electrolyte interface (a) in the absence of a co-catalyst (b) in the presence of a surface passivating CoPi cocatalyst and (c) in the presence of a non-passivating RuOₓ. The three possible routes for the photogenerated holes movement is described. Firstly, the movement of the photogenerated hole towards the semiconductor-electrolyte interface for the direct water oxidation reaction. Secondly, the surface state-mediated recombination. And lastly, the charge transfer through the surface co-catalyst for water oxidation reaction kₜᵣ Refers to the charge transfer rate and kᵣₑ꜀ refers to surface recombination [164].

Another critical factor to be considered in $BiVO_4$ photoanode is the effect on photovoltage with or without the surface co-catalyst in the presence and absence of the illumination. Because the introduction of a passivating cocatalyst layer reduces the potential loss due to the reaction kinetics in the surface Helmholtz layer, therefore, a significant change in photovoltage is expected in the dark and under the irradiation. Experimentally, the photovoltage of only $BiVO_4$ photoanode (without surface passivating layer) was measured to be 0.16 V which was noticed to be increased to 0.28 V upon the surface passivation co-catalyst such as CoOOH. This clearly confirms that the photovoltage of BiVO$_4$ electrode could be increased dramatically by deactivating the surface trap states [151]. When BiVO$_4$ photoanode comes in contact with the electrolyte, the alignment of Fermi level of BiVO$_4$ and the redox potential level of the electrolyte takes place causing band bending at the interface. Fig. 11a shows the band diagram of BiVO$_4$ showing the photoresponse in the presence and absence of illumination. Under dark and without the surface passivating layer, the open-circuit voltage of BiVO$_4$ depends on its carrier density and the redox potential energy level of the electrolyte. However, during illumination, the generation and transportation of carriers to the BiVO$_4$ surface alter the electron density at the surface. Based on the carrier density of BiVO$_4$ at the surface, a new quasi-Fermi level is introduced. Under such circumstances, the presence of surface states at the BiVO$_4$ surface could trap the partially photogenerated carriers causing a significant loss in the surface carrier density which ultimately decreases the photovoltage. After passivating the BiVO$_4$ surface with CoOOH, the density of surface trap states decreases. Hence, the use of a passivation layer shows a significant increase in the photovoltage and consequently reduces the water oxidation potential and enhances the PEC activity.

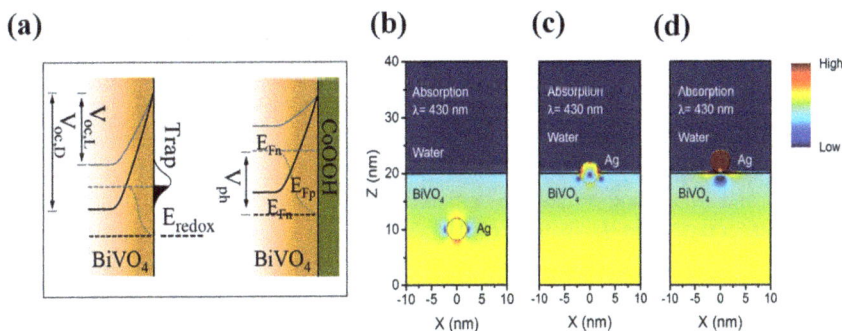

Figure 11: (a) Surface band edge diagram of BiVO$_4$ in the presence and absence of surface passivation layer showing the photovoltage response under dark (black lines) and illumination (grey lines) conditions. [151] (Reprinted with permission from American Chemical Society). The absorption and distribution of light in a system containing single Ag nanoparticles in (b) half-embedded, (c) impregnated, and (d) surface-attached configuration with BiVO$_4$. The simulations were based on the light irradiation of 430 nm wavelength and 4nm of spherical Ag nanoparticles [162] (Reprinted with permission from American Chemical Society).

The band gap of $BiVO_4$ allows the absorption of a limited amount of incident light which confines its water splitting efficiency. This limitation can be solved by sensitizing the $BiVO_4$ with the narrow bandgap materials possessing suitable band edge positions. Some of the narrow bandgap sensitizers used with $BiVO_4$ and their PEC water splitting properties are included in Table 1. That is, the effective sensitizers such as carbon quantum dot [123], Ag_2S [131], Fe_2O_3 [135], Bi_2S_3[138,155], Fe_2TiO_5 [143] have been successfully used to improve the electrochemical properties of $BiVO_4$ photoanode. Including sensitizers mentioned above, the use of a surface catalyst is still necessary to minimize the charge recombination and to achieve significantly high PEC water splitting efficiency. Among the combination of co-catalyst and sensitizers used so far, the $BiVO_4$/carbon quantum dot/FeOOH/NiOOH composite electrode exhibited remarkable photocurrent of 5.99 mA cm^{-2} (at 1.23 V vs RHE) [123]. Importantly, such a high photocurrent was achieved without the use of any hole scavenger in KH_2PO_4 aqueous electrolyte of pH = 7. In this composite, the use of carbon quantum dot increases the absorption range of $BiVO_4$ in the entire visible region, whereas the surface passivation catalyst FeOOH/NiOOH improves the water oxidation kinetics by suppressing the carrier recombination [123] yielding 2.29% of applied bias photon-to-current efficiency. Furthermore, with long-time stability of 10 h of PEC operation and 95% of Faradaic efficiency, an ~80 and ~40 μmol h^{-1} cm^{-2} of hydrogen and oxygen were generated, respectively.

Recently, Gao et al. [153] have proposed $BiVO_4$/NiMoO$_4$/CoPi photoanodes and studied their PEC water splitting activity. In the composite, the NiMoO$_4$ extends the light absorption ability and charge separation efficiency during illumination. The surface catalysis and the photoconversion efficiency were improved by passivating the $BiVO_4$/NiMoO$_4$ electrode using Co-Pi catalyst. With the optimized electrode configuration, about 5.3 mA cm^{-2} of current under illumination and 1.18% photoconversion efficiency were recorded at 1.23 and 0.86 V vs RHE, respectively [153]. Similarly, various materials and catalysts utilized for enhancing the PEC activity of $BiVO_4$ photoanode are shown in Table 1. However, we discussed only a few of the best performing photoanode combinations above as the underlying mechanism of photocurrent enhancement remains the same for all the catalyst and sensitizer supported $BiVO_4$ photoanodes. All the variety of utilized passivation co-catalyst layers shown in Table 1 served the same purpose as discussed above, that is suppression charge recombination and improving the kinetics of water oxidation.

Table 1. PEC water splitting performances of BiVO$_4$ photoanodes

Electrode configuration	Photocurrent (mA cm^2)	IPCE (%)	H$_2$ and O$_2$ gas evolution rates	Ref.
BiVO$_4$/CoO$_x$/NiO	2.5 at 0.6 V vs. RHE	~45 at 420 nm	H$_2$: ~37.5 µmol h^{-1} cm^{-2} O$_2$: ~18.7 µmol h^{-1} cm^{-2}	[114]
Mo:BiVO$_4$/Au	3.1 at 1.23 V vs. RHE	~40 at 400 nm	H$_2$: 7.7 µmol h^{-1} cm^{-2} O$_2$: 23.8 µmol h^{-1} cm^{-2}	[115]
BiVO$_4$/Co-borate	2.5 at 0.7 V vs. SCE	~45 at 400 nm	H$_2$: ~23 µmol h^{-1} cm^{-2} O$_2$: ~12 µmol h^{-1} cm^{-2}	[116]
W:BiVO$_4$/Co-Pi	1.5 at 1.23 V vs. RHE	~33 at 400 nm	O$_2$: ~ ~2.4 nmol s^{-1} cm^{-2}	[12]
BiVO$_4$/FeOOH/NiOOH	2.73 at 0.6 V vs. RHE	~72 at 420 nm	H$_2$: ~3.58 µmol h^{-1} cm^{-2} O$_2$: ~1.83 µmol h^{-1} cm^{-2}	[117]
BiVO$_4$/PdO$_x$	1.15 at 1.23 V vs. RHE	~25 at 400 nm	-	[118]
BiVO$_4$/NiO$_x$	1.57 at 1.23 V vs. RHE	~43 at 450 nm	-	[119]
Mo:BiVO$_4$/FeOOH	3.5 at 1.23 V vs. RHE	-	-	[120]
BiVO$_4$/ Ru-based molecular catalyst	~1.4 at 1.23 V vs. RHE	-	-	[121]
BiVO$_4$/Au	2.4 at 1.23 V vs. RHE	~50 at 450 nm	-	[122]
BiVO$_4$/Carbon Quantum Dot /FeOOH/NiOOH	5.99 at 1.23 V vs. RHE	~72 at 400 nm	H$_2$: ~80 µmol h^{-1} cm^{-2} O$_2$: ~40 µmol h^{-1} cm^{-2}	[123]
Mo:BiVO$_4$	1.78 at 1.2 V vs. Ag/AgCl	-	-	[124]
BiVO$_4$/Cobalt–Iron hydroxide (CoFe-H)	2.48 at 1.23 V vs. RHE	-	H$_2$: ~18 µmol h^{-1} cm^{-2} O$_2$: ~8 µmol h^{-1} cm^{-2}	[11]
BiVO$_4$/CoAl-LDH/graphene	2.13 at 1.23 V vs. RHE	~60 at 360 nm	O$_2$: ~8 µmol h^{-1} cm^{-2}	[125]
BiVO$_4$/CoFe-PB	0.95 at 1.23 V vs. RHE	~33 at 400 nm	O$_2$: ~3 µmol h^{-1} cm^{-2}	[126]
BiVO$_4$/Co$_4$O$_4$ cubane	5 at 1.23 V vs. RHE	~93 at 420 nm	O$_2$: ~20 nmol h^{-1} cm^{-2}	[127]
BiVO$_4$/Co−La mixed LDH	~2 at 1.23 V vs. RHE	-	O$_2$: ~10 nmol h^{-1} cm^{-2}	[128]
Mo:meso-BiVO$_4$/Co-Pi	4.57 at 1.23 V vs. RHE	~70 at 360 nm	-	[129]
BiVO$_4$ nanopillars	1.23 at 1.2 V vs. A/AgCl	-	-	[130]
BiVO$_4$/Ag$_2$S	1.91 at 1.23 V vs. RHE	~23 at 460 nm	-	[131]

$BiVO_4/rGO/NiFe$	1.30 at 1.23 V vs. RHE	~25 at 350 nm	-	[132]
$Mo:BiVO_4/NiO_x$	2.44 at 1.23 V vs. RHE	~50 at 350 nm	-	[133]
$BiVO_4/Bi$	1.96 at 1.23 V vs. RHE	~70 at 420 nm	-	[134]
$BiVO_4/Fe_2O_3/FeOOH/$ NiOOH	2.24 at 1.23 V vs. RHE	~37 at 400 nm	H_2: ~1.25 µmol h^{-1} cm^{-2} O_2: ~0.7 µmol h^{-1} cm^{-2}	[135]
$BiVO_4/ZnO/TiO_2$	2.25 at 1.2 V vs. A/AgCl	~65 at 350 nm	-	[136]
$Zr:BiVO_4/Ag_3PO_4$	2.3 at 1.23 V vs. RHE	~35 at 350 nm	O_2: ~16 µmol h^{-1} cm^{-2}	[137]
$BiVO_4/Bi_2S_3/NiOOH$	0.91 at 1.23 V vs. RHE	~5.5 at 200 nm	-	[138]
$BiVO_4/FeO_x$	~1.1 at 1.23 V vs. RHE	~20 at 400 nm	-	[139]
$BiVO_4/FeCoO_x$	4.82 at 1.23 V vs. RHE	~86 at 400 nm	H_2: ~86.5 µmol h^{-1} cm^{-2} O_2: ~42.0 µmol h^{-1} cm^{-2}	[140]
$BiVO_4/TiO_2/Ir-$ $COOH/TiO_2$	3.3 at 1.23 V vs. RHE	~48 at 450 nm	-	[141]
$BiVO_4/NiFe-LDH/CDs$	2.84 at 1.23 V vs. RHE	~40 at 350 nm	-	[142]
$BiVO_4/Fe_2TiO_5$	3.2 at 1.23 V vs. RHE	-	-	[143]
$BiVO_4/Mn-Bi$	0.94 at 1.5 V s. RHE	-	-	[144]
$BiVO_4/FeF_2$	2.49 at 1.23 V vs. RHE	~28 at 450 nm	H_2: ~280 µmol h^{-1} cm^{-2}	[145]
$BiVO_4$	2.3 at 1.23 V vs. RHE	-	-	[146]
$BiVO_4/poly[-$ $Co_2(benzimidazole)_4]$	3.1 at 1.23 V vs. RHE	~50 at 420 nm	-	[147]
$BiVO_4/cobalt\ salophen$	4.27 at 1.23 V vs. RHE	~89 at 430 nm	O_2: ~27.78 µmol h^{-1} cm^{-2}	[148]
$BiVO_4/AgO_x/NiO_x$	2.05 at 1.23 V vs. RHE	~49.3 at 420 nm	H_2: ~8 µmol h^{-1} cm^{-2} O_2: ~4 µmol h^{-1} cm^{-2}	[149]
$BiVO_4/Ag/MoS_2$	2.72 at 1.23 V vs. RHE	~51 at 420 nm	H_2: ~57.8 µmol h^{-1} cm^{-2} O_2: ~29.7 µmol h^{-1} cm^{-2}	[150]
$BiVO_4/\ CoOOH$	4.0 at 1.23 V vs. RHE	~65 at 450 nm	-	[151]
$BiVO_4/PANI/NiOOH$	3.31 at 1.23 V vs. RHE	~83.3 at 450 nm	H_2: ~30 µmol h^{-1} cm^{-2} O_2: ~15 µmol h^{-1} cm^{-2}	[152]
$BiVO_4/NiMoO_4/Co-Pi$	5.3 at 1.23 V vs. RHE	~55 at 450 nm	H_2: ~2.8 µmol h^{-1} cm^{-2} O_2: ~1.3 µmol h^{-1} cm^{-2}	[153]
$Mo:BiVO_4/RGO$	8.51 at 1.23 V vs. RHE	~25 at 420 nm	H_2: ~1.43 µmol h^{-1} cm^{-2}	[154]

$BiVO_4/Bi_2S_3/FeOOH$	0.8 at 1.23 V vs. RHE	-	-	[155]
$BiVO_4/TiO_2$/Cobalt borate/TiO_2	2.5 at 1.23 V vs. RHE	~45 at 450 nm	H_2: ~30 µmol h^{-1} cm^{-2} O_2: ~14 µmol h^{-1} cm^{-2}	[156]
$BiVO_4$/CoMnZn LDH	1.06 at 1.23 V vs. RHE	~55 at 350 nm	O_2: ~7.5 µmol h^{-1} cm^{-2}	[157]
$BiVO_4$/Ag/N-doped carbon	2.42 at 1.23 V vs. RHE	~30 at 400 nm	-	[158]
Zr:$BiVO_4$/Ni-Fe/prussian blue polymer	3.23 at 1.23 V vs. RHE	~50 at 460 nm	O_2: ~9 µmol h^{-1} cm^{-2}	[159]
$BiVO_4/Al_2O_3$	3.0 at 1.23 V vs. RHE	~60 at 350 nm	-	[160]
$BiVO_4$/ferrite phosphate	2.28 at 1.23 V vs. RHE	~50 at 450 nm	-	[161]
$BiVO_4$/Ag	1.7 at 1.23 V vs. RHE	~20 at 450 nm	-	[162]
$BiVO_4$/Au	1.1 at 0.6 V vs. Ag/AgCl	~32.5 at 440 nm	H_2: ~12.5 µmol h^{-1} cm^{-2} O_2: ~7 µmol h^{-1} cm^{-2}	[163]

13. Plasmon-induced resonant energy transfer

The limited visible light absorption ability of $BiVO_4$ can also be extended by utilizing the localized surface plasmon resonance (LSPR) generated from the noble metal nanoparticles attached to the $BiVO_4$. The creation of LSPR in metal nanoparticles is the result of collective oscillation of conduction band electrons excited via incident electromagnetic radiation. Generally, the improved PEC activity in LSPR exhibiting metal nanoparticle embedded in $BiVO_4$ has been ascribed to the light absorption enhancement and plasmon-enhanced energy transfer. Using the LSPR, the absorption range of metal nanoparticles can be tailored in the visible region based on the peak resonant wavelength adjusted by tuning the size and shape of metal nanoparticles. [165,166]. Hence, the metal oxide semiconductors sensitized with metal nanoparticles show enhanced PEC properties. Here, the resonant wavelength of metal nanoparticle should match well with the absorption edge of the metal oxide material for efficient utilization of LSPR properties. During the illumination on the metal nanoparticle embedded $BiVO_4$, the efficient use of LSPR depends on the influence of LSPR-induced electric field penetration depth. Therefore, the nanostructure arrangement of LSPR materials on $BiVO_4$ is important. The LSPR exhibiting metal nanoparticles such as Au [115] and Ag [162] are routinely used to enhance PEC properties of $BiVO_4$ photoanodes. Notably, Ag nanoparticle was reported to be suitable with its 360 to 500 nm of the resonant wavelength as it matches well with the $BiVO_4$ absorption onset values [162].

Hence, the LSPR mediated effect of Ag on $BiVO_4$ exhibits efficient generation of carriers and effective separation of charges yielding higher photocurrents and lower onset potentials [162].

As mentioned earlier, due to the low and limited diffusion of the LSPR-induced electric field, the carrier generation in $BiVO_4$ finitely occurs in the vicinity of metal nanoparticles. Therefore, for the efficient PEC water splitting, the maximum exploitation of LSPR-induced effect has been highly essential. [162]. Fig. 11b shows the simulation studies depicting absorption and distribution light taking place on $BiVO_4$ impregnated, half-embedded, and surface-attached single Ag nanoparticles. Conclusively, the Ag nanoparticles in impregnated mode successfully help the absorption of light in the bulk of $BiVO_4$. In contrast, the configuration containing half embedded and surface-attached Ag nanoparticles affect only the near surface region. Undeniably, the experimental results show that Ag nanoparticles impregnated in $BiVO_4$ enhanced PEC activity due to the enhanced carrier generation and separation kinetics [162].

The shape of the LSPR metal nanoparticles (embedded on $BiVO_4$) is also essential for achieving higher PEC activity as demonstrated by Lee et al., [122]. Among the hemispherical and octahedral Au nanoparticles, the sharp vertices and well-defined edges possessed by octahedral Au nanoparticles offer comparatively more intensification of the local field. Hence, the octahedral Au nanoparticles show a more efficient oxygen evolution reaction compared to the hemispherical shape Au nanoparticles. Notably, the octahedral nanoparticles were found to offer higher absorption and efficient charge generation and separation via LSPR induced effects [122]. Furthermore, the morphology of $BiVO_4$, particle sizes of Au [163] and their distribution could also influence the PEC activity. For example, the Au nanoparticles attached to the inverse opal structure of $BiVO_4$ effectively manipulate and confine light absorption yielding efficient charge carrier generation and separation [115]. Such a photonic nanostructure (Au nanoparticles attached to the inverse opal structure of $BiVO_4$) enhances the surface plasmon resonance via photonic Bragg resonance [115]. In the $Au/BiVO_4$ system, the hot spots between the $BiVO_4$ can be generated by Au nanoparticles by tuning its size, pitch and patterning. That is, by hexagonal patterning and tuning Au nanoparticles on $BiVO_4$ plasmonic hot spots could be generated which could enhance the light absorption capabilities (over the ultraviolet-visible region) [129]. Via such nanostructure, the efficient charge transport and transfer could be achieved [129]. It was revealed that the patterned nanosphere Au enhances the lifetime of both electrons and holes due to plasmon-induced resonant energy transfer.

Conclusions and future perspective

The discussion of this chapter concludes that $BiVO_4$ indeed offers an efficient and sustainable PEC water splitting to produce both O_2 and H_2. The band-gap and the suitable band edge potentials of $BiVO_4$ allow effective utilization as a photoanode component to perform overall water splitting under illumination. Since last few years, the research interest on $BiVO_4$ solar water splitting has tremendously increased indicating its promising features in future solar water splitting systems. The greater theoretical studies supported with experimental results has helped in engineering the $BiVO_4$ with optimized optical, electrical and catalytic properties for the enhanced water splitting activity. Based on the recent theoretical and experimental results, the factors influencing, limiting and affecting the water splitting activity are briefed and discussed. The factors such as the crystal structure, facet, pH of the electrolyte, carrier dynamics, intrinsic defects, oxygen vacancies, doping, polarons, plasmon-induced resonant energy transfer and surface passivation catalysts are known to cause serious influence on the PEC water splitting efficiency of $BiVO_4$. A significant milestone has been achieved in producing photocurrent, IPCE, photoconversion efficiency, and H_2/O_2 production via controlling and optimizing facets, electrolyte pH, oxygen vacancies, defects, catalyst, dopant, carrier dynamics, polarons, and surface passivation catalyst. The key factors and their influence discussed in this chapter are expected to offer a thorough understanding of the present state of $BiVO_4$ photoanode in PEC water splitting system. The key findings in this chapter indicate that few areas of $BiVO_4$ still needs optimization for its much more efficient utilization in PEC water splitting system. That is, the large scale/size production of $BiVO_4$ photoanode and PEC systems needs to be engineered, and structure and surface modifications in the direction of longtime water splitting stability needs to be addressed. By solving these issues could significantly benefit and direct the researchers to focus more on commercialization of the PEC technology. Such breakthroughs will provide a milestone in sustainable production of clean fuel technology through solar water splitting.

References

[1]   A. Fujishima, K. Honda, Electrochemical photolysis of water at a semiconductor electrode, Nature 238 (1972) 238037a0. https://doi.org/10.1038/238037a0.

[2]   A. Kudo, K. Ueda, H. Kato, I. Mikami, Photocatalytic O_2 evolution under visible light irradiation on $BiVO_4$ in aqueous $AgNO_3$ solution, Catal. Lett. 53 (1998) 229–230. https://doi.org/10.1023/A:1019034728816.

[3]   Y. Park, K. J. McDonald, K.-S. Choi, Progress in bismuth vanadate photoanodes for use in solar water oxidation, Chem. Soc. Rev. 42 (2013) 2321–2337. https://doi.org/10.1039/C2CS35260E.

[4]   B. Xie, H. Zhang, P. Cai, R. Qiu, Y. Xiong, Simultaneous photocatalytic reduction of Cr(VI) and oxidation of phenol over monoclinic $BiVO_4$ under visible light irradiation, Chemosphere 63 (2006) 956–963. https://doi.org/10.1016/j.chemosphere.2005.08.064.

[5]   M.S. Prévot, K. Sivula, Photoelectrochemical Tandem Cells for Solar Water Splitting, J. Phys. Chem. C. 117 (2013) 17879–17893. https://doi.org/10.1021/jp405291g.

[6]   S.R. Taylor, Abundance of chemical elements in the continental crust: A new table, Geochim. Cosmochim. Acta 28 (1964) 1273–1285. https://doi.org/10.1016/0016-7037(64)90129-2.

[7]   K.R. Tolod, S. Hernández, N. Russo, Recent advances in the $BiVO_4$ photocatalyst for sun-driven water oxidation: top-performing photoanodes and scale-up challenges, Catalysts 7 (2017) 13. https://doi.org/10.3390/catal7010013.

[8]   H.D. Telpande, D.V. Parwate, Characterization supported improved method for the synthesis of bismuth vanadate and its assessment with conventional synthetic route, IOSR J. Appl. Chem. 8 (2015) 28-37.

[9]   A.R. Lim, S.H. Choh, M.S. Jang, Prominent ferroelastic domain walls in $BiVO_4$ crystal, J. Phys. Condens. Matter. 7 (1995) 7309–7323. https://doi.org/10.1088/0953-8984/7/37/005.

[10]   F.F. Abdi, N. Firet, R. van de Krol, Efficient $BiVO_4$ thin film photoanodes modified with cobalt phosphate catalyst and W-doping, ChemCatChem 5 (2013) 490–496. https://doi.org/10.1002/cctc.201200472.

[11]   W. Liu, H. Liu, L. Dang, H. Zhang, X. Wu, B. Yang, Z. Li, X. Zhang, L. Lei, S. Jin, Amorphous cobalt–iron hydroxide nanosheet electrocatalyst for efficient electrochemical and photo-electrochemical oxygen evolution, Adv. Funct. Mater. 27 (2017) 1603904. https://doi.org/10.1002/adfm.201603904.

[12]   D.K. Zhong, S. Choi, D.R. Gamelin, Near-complete suppression of surface recombination in solar photoelectrolysis by "Co-Pi" catalyst-modified W:$BiVO_4$, 133 (2011) 18370-18377. https://doi.org/10.1021/ja207348x.

[13]   High Performance Pigments, 1st ed., John Wiley & Sons, Ltd, 2002. https://doi.org/10.1002/3527600493.

[14]   A. Kudo, K. Omori, H. Kato, A novel aqueous process for preparation of crystal form-controlled and highly crystalline $BiVO_4$ powder from layered vanadates at room temperature and its photocatalytic and photophysical properties, J. Am. Chem. Soc. 121 (1999) 11459–11467. https://doi.org/10.1021/ja992541y.

[15]   A.W. Sleight, H. -y. Chen, A. Ferretti, D.E. Cox, Crystal growth and structure of $BiVO_4$, Mater. Res. Bull. 14 (1979) 1571–1581. https://doi.org/10.1016/0025-5408(72)90227-9.

[16]   A.K. Bhattacharya, K.K. Mallick, A. Hartridge, Phase transition in $BiVO_4$, Mater. Lett. 30 (1997) 7–13. https://doi.org/10.1016/S0167-577X(96)00162-0.

[17]   C. Martinez Suarez, S. Hernández, N. Russo, $BiVO_4$ as photocatalyst for solar fuels production through water splitting: A short review, Appl. Catal. Gen. 504 (2015) 158–170. https://doi.org/10.1016/j.apcata.2014.11.044.

[18]   J.K. Cooper, S. Gul, F.M. Toma, L. Chen, P.-A. Glans, J. Guo, J.W. Ager, J. Yano, I.D. Sharp, Electronic Structure of Monoclinic $BiVO_4$, Chem. Mater. 26 (2014) 5365–5373. https://doi.org/10.1021/cm5025074.

[19]   Z.-F. Huang, L. Pan, J.-J. Zou, X. Zhang, L. Wang, Nanostructured bismuth vanadate-based materials for solar-energy-driven water oxidation: a review on recent progress, Nanoscale. 6 (2014) 14044–14063. https://doi.org/10.1039/C4NR05245E.

[20]   A. Walsh, Y. Yan, M.N. Huda, M.M. Al-Jassim, S.-H. Wei, Band Edge Electronic Structure of $BiVO_4$: Elucidating the role of the Bi s and V d Orbitals, Chem. Mater. 21 (2009) 547–551. https://doi.org/10.1021/cm802894z.

[21]   K. Ding, B. Chen, Z. Fang, Y. Zhang, Density functional theory study on the electronic and optical properties of three crystalline phases of $BiVO_4$, Theor. Chem. Acc. 132 (2013) 1352. https://doi.org/10.1007/s00214-013-1352-x.

[22]   Z. Zhao, Z. Li, Z. Zou, Electronic structure and optical properties of monoclinic clinobisvanite $BiVO_4$, Phys. Chem. Chem. Phys. 13 (2011) 4746–4753. https://doi.org/10.1039/C0CP01871F.

[23]   J. Ma, L.W. Wang, The role of the isolated 6s states in $BiVO_4$ on the electronic and atomic structures, Appl. Phys. Lett. 105 (2014) 172102. https://doi.org/10.1063/1.4900549.

[24]   S. Stoughton, M. Showak, Q. Mao, P. Koirala, D.A. Hillsberry, S. Sallis, L.F. Kourkoutis, K. Nguyen, L.F.J. Piper, D.A. Tenne, N.J. Podraza, D.A. Muller, C. Adamo, D.G. Schlom, Adsorption-controlled growth of $BiVO_4$ by molecular-beam epitaxy, APL Mater. 1 (2013) 042112. https://doi.org/10.1063/1.4824041.

[25]   J.K. Cooper, S. Gul, F.M. Toma, L. Chen, Y.-S. Liu, J. Guo, J.W. Ager, J. Yano, I.D. Sharp, Indirect Bandgap and Optical Properties of Monoclinic Bismuth Vanadate, J. Phys. Chem. C. 119 (2015) 2969–2974. https://doi.org/10.1021/jp512169w.

[26]   F. Ambrosio, G. Miceli, A. Pasquarello, Redox levels in aqueous solution: Effect of van der Waals interactions and hybrid functionals, J. Chem. Phys. 143 (2015) 244508. https://doi.org/10.1063/1.4938189.

[27]   F. Ambrosio, J. Wiktor, A. Pasquarello, pH-dependent catalytic reaction pathway for water splitting at the $BiVO_4$–water interface from the band alignment, ACS Energy Lett. 3 (2018) 829–834. https://doi.org/10.1021/acsenergylett.8b00104.

[28]   R. Li, F. Zhang, D. Wang, J. Yang, M. Li, J. Zhu, X. Zhou, H. Han, C. Li, Spatial separation of photogenerated electrons and holes among {010} and {110} crystal facets of $BiVO_4$, Nat. Commun. 4 (2013) 1432. https://doi.org/10.1038/ncomms2401.

[29]   T. Tachikawa, T. Ochi, Y. Kobori, Crystal-face-dependent charge dynamics on a $BiVO_4$ photocatalyst revealed by single-particle spectroelectrochemistry, ACS Catal. 6 (2016) 2250–2256. https://doi.org/10.1021/acscatal.6b00234.

[30]   C.W. Kim, Y.S. Son, M.J. Kang, D.Y. Kim, Y.S. Kang, (040)-Crystal facet engineering of $BiVO_4$ plate photoanodes for solar fuel production, Adv. Energy Mater. 6 (2016) 1501754. https://doi.org/10.1002/aenm.201501754.

[31]   J. Song, M.J. Seo, T.H. Lee, Y.-R. Jo, J. Lee, T.L. Kim, S.-Y. Kim, S.-M. Kim, S.Y. Jeong, H. An, S. Kim, B.H. Lee, D. Lee, H.W. Jang, B.-J. Kim, S. Lee, Tailoring crystallographic orientations to substantially enhance charge separation efficiency in anisotropic $BiVO_4$ photoanodes, ACS Catal. 8 (2018) 5952–5962. https://doi.org/10.1021/acscatal.8b00877.

[32]   J. Hu, W. Chen, X. Zhao, H. Su, Z. Chen, Anisotropic electronic characteristics, adsorption, and stability of low-index $BiVO_4$ surfaces for photoelectrochemical applications, ACS Appl. Mater. Interfaces 10 (2018) 5475–5484. https://doi.org/10.1021/acsami.7b15243.

[33]   A.J.E. Rettie, W.D. Chemelewski, J. Lindemuth, J.S. McCloy, L.G. Marshall, J. Zhou, D. Emin, C.B. Mullins, Anisotropic small-polaron hopping in $W:BiVO_4$ single crystals, Appl. Phys. Lett. 106 (2015) 022106. https://doi.org/10.1063/1.4905786.

[34]   H.S. Han, S. Shin, D.H. Kim, I.J. Park, J.S. Kim, P.-S. Huang, J.-K. Lee, I.S. Cho, X. Zheng, Boosting the solar water oxidation performance of a $BiVO_4$ photoanode by crystallographic orientation control, Energy Environ. Sci. 11 (2018) 1299–1306. https://doi.org/10.1039/C8EE00125A.

[35]   J. Yang, D. Wang, X. Zhou, C. Li, A theoretical study on the mechanism of photocatalytic oxygen evolution on $BiVO_4$ in aqueous solution, Chem. Eur. J. 19 (2013) 1320–1326. https://doi.org/10.1002/chem.201202365.

[36]   L. Zhang, W. Niu, G. Xu, Synthesis and applications of noble metal nanocrystals with high-energy facets, Nano Today 7 (2012) 586–605. https://doi.org/10.1016/j.nantod.2012.10.005.

[37]   Q. Kuang, X. Wang, Z. Jiang, Z. Xie, L. Zheng, High-energy-surface engineered metal oxide micro- and nanocrystallites and their applications, Acc. Chem. Res. 47 (2014) 308–318. https://doi.org/10.1021/ar400092x.

[38]   Z. Quan, Y. Wang, J. Fang, High-index faceted noble metal nanocrystals, Acc. Chem. Res. 46 (2013) 191–202. https://doi.org/10.1021/ar200293n.

[39]   P. Li, X. Chen, H. He, X. Zhou, Y. Zhou, Z. Zou, Polyhedral 30-faceted $BiVO_4$ microcrystals predominantly enclosed by high-index planes promoting photocatalytic water-splitting activity, Adv. Mater. 30 (2018) 1703119. https://doi.org/10.1002/adma.201703119.

[40]   Y. Ma, S. R. Pendlebury, A. Reynal, F.L. Formal, J. R. Durrant, Dynamics of photogenerated holes in undoped $BiVO_4$ photoanodes for solar water oxidation, Chem. Sci. 5 (2014) 2964–2973. https://doi.org/10.1039/C4SC00469H.

[41]   J. Ravensbergen, F.F. Abdi, J.H. van Santen, R.N. Frese, B. Dam, R. van de Krol, J.T.M. Kennis, Unraveling the carrier dynamics of $BiVO_4$: A femtosecond to microsecond transient absorption study, J. Phys. Chem. C 118 (2014) 27793–27800. https://doi.org/10.1021/jp509930s.

[42]   M. Yabuta, A. Takeda, T. Sugimoto, K. Watanabe, A. Kudo, Y. Matsumoto, Particle size dependence of carrier dynamics and reactivity of photocatalyst $BiVO_4$ probed with single-particle transient absorption microscopy, J. Phys. Chem. C 121 (2017) 22060–22066. https://doi.org/10.1021/acs.jpcc.7b06230.

[43]   H. Inoue, T. Shimada, Y. Kou, Y. Nabetani, D. Masui, S. Takagi, H. Tachibana, The water oxidation bottleneck in artificial photosynthesis: How can we get through it? an alternative route involving a two-electron process, ChemSusChem 4 (2011) 173–179. https://doi.org/10.1002/cssc.201000385.

[44]   Z.-Y. Zhao, Single water molecule adsorption and decomposition on the low-index stoichiometric rutile TiO_2 surfaces, J. Phys. Chem. C 118 (2014) 4287–4295. https://doi.org/10.1021/jp500177n.

[45]   S. Lardhi, L. Cavallo, M. Harb, Determination of the intrinsic defect at the origin of poor H_2 evolution performance of the monoclinic $BiVO_4$ photocatalyst using density functional theory, J. Phys. Chem. C. 122 (2018) 18204–18211. https://doi.org/10.1021/acs.jpcc.8b03044.

[46]   J. Hu, X. Zhao, W. Chen, H. Su, Z. Chen, Theoretical insight into the mechanism of photoelectrochemical oxygen evolution reaction on $BiVO_4$ anode with oxygen vacancy, J. Phys. Chem. C 121 (2017) 18702–18709. https://doi.org/10.1021/acs.jpcc.7b05884.

[47]   M. Lamers, S. Fiechter, D. Friedrich, F.F. Abdi, R. van de Krol, Formation and suppression of defects during heat treatment of $BiVO_4$ photoanodes for solar water splitting, J. Mater. Chem. A. 6 (2018) 18694–18700. https://doi.org/10.1039/C8TA06269B.

[48]   T.W. Kim, Y. Ping, G.A. Galli, K.-S. Choi, Simultaneous enhancements in photon absorption and charge transport of bismuth vanadate photoanodes for solar water splitting, Nat. Commun. 6 (2015) 8769. https://doi.org/10.1038/ncomms9769.

[49]   L. Zhou, C. Zhao, B. Giri, P. Allen, X. Xu, H. Joshi, Y. Fan, L.V. Titova, P.M. Rao, High light absorption and charge separation efficiency at low applied voltage from Sb-Doped SnO_2/$BiVO_4$ core/shell nanorod-array photoanodes, Nano Lett. 16 (2016) 3463–3474. https://doi.org/10.1021/acs.nanolett.5b05200.

[50]   J.H. Kim, J.-W. Jang, Y.H. Jo, F.F. Abdi, Y.H. Lee, R. van de Krol, J.S. Lee, Hetero-type dual photoanodes for unbiased solar water splitting with extended light harvesting, Nat. Commun. 7 (2016) 13380. https://doi.org/10.1038/ncomms13380.

[51]   S. Wang, P. Chen, J.-H. Yun, Y. Hu, L. Wang, An electrochemically treated $BiVO_4$ photoanode for efficient photoelectrochemical water splitting, Angew. Chem. Int. Ed. 56 (2017) 8500–8504. https://doi.org/10.1002/anie.201703491.

[52]   J.-M. Wu, Y. Chen, L. Pan, P. Wang, Y. Cui, D. Kong, L. Wang, X. Zhang, J.J. Zou, Multi-layer monoclinic $BiVO_4$ with oxygen vacancies and V^{4+} species for highly efficient visible-light photoelectrochemical applications, Appl. Catal. B Environ. 221 (2018) 187–195. https://doi.org/10.1016/j.apcatb.2017.09.031.

[53]   S. Wang, P. Chen, Y. Bai, J.-H. Yun, G. Liu, L. Wang, New $BiVO_4$ dual photoanodes with enriched oxygen vacancies for efficient solar-driven water splitting, Adv. Mater. 30 (2018) 1800486. https://doi.org/10.1002/adma.201800486.

[54]   Y. Zhang, D. Wang, X. Zhang, Y. Chen, L. Kong, P. Chen, Y. Wang, C. Wang, L. Wang, Y. Liu, Enhanced photoelectrochemical performance of nanoporous $BiVO_4$

photoanode by combining surface deposited cobalt-phosphate with hydrogenation treatment, Electrochim. Acta 195 (2016) 51–58. https://doi.org/10.1016/j.electacta.2016.02.137.

[55]   J. Li, L. Guo, N. Lei, Q. Song, Z. Liang, Metallic Bi Nanocrystal-Modified Defective BiVO$_4$ Photoanodes with Exposed (040) Facets for Photoelectrochemical Water Splitting, ChemElectroChem 4 (2017) 2852–2861. https://doi.org/10.1002/celc.201700680.

[56]   M.D. Bhatt, J.Y. Lee, Photocatalytic properties of intrinsically defective undoped bismuth vanadate (BiVO$_4$) photocatalyst: A DFT study, J. Electroanal. Chem. 828 (2018) 97–101. https://doi.org/10.1016/j.jelechem.2018.09.042.

[57]   D. Emin, Polarons by David Emin, Camb. Core. (2012). https://doi.org/10.1017/CBO9781139023436.

[58]   M. Ziwritsch, S. Müller, H. Hempel, T. Unold, F.F. Abdi, R. van de Krol, D. Friedrich, R. Eichberger, Direct time-resolved observation of carrier trapping and polaron conductivity in BiVO$_4$, ACS Energy Lett. 1 (2016) 888–894. https://doi.org/10.1021/acsenergylett.6b00423.

[59]   J. Wiktor, F. Ambrosio, A. Pasquarello, Role of polarons in water splitting: The case of BiVO$_4$, ACS Energy Lett. 3 (2018) 1693–1697. https://doi.org/10.1021/acsenergylett.8b00938.

[60]   A.J.E. Rettie, W.D. Chemelewski, D. Emin, C.B. Mullins, Unravelling small-polaron transport in metal oxide photoelectrodes, J. Phys. Chem. Lett. 7 (2016) 471–479. https://doi.org/10.1021/acs.jpclett.5b02143.

[61]   K.E. Kweon, G.S. Hwang, J. Kim, S. Kim, S. Kim, Electron small polarons and their transport in bismuth vanadate: a first principles study, Phys. Chem. Chem. Phys. 17 (2014) 256–260. https://doi.org/10.1039/C4CP03666B.

[62]   V. Jovic, J. Laverock, A.J.E. Rettie, J.-S. Zhou, C.B. Mullins, V.R. Singh, B. Lamoureux, D. Wilson, T.-Y. Su, B. Jovic, H. Bluhm, T. Söhnel, K.E. Smith, Soft X-ray spectroscopic studies of the electronic structure of M:BiVO$_4$ (M = Mo, W) single crystals, J. Mater. Chem. A. 3 (2015) 23743–23753. https://doi.org/10.1039/C5TA07898A.

[63]   F.F. Abdi, T.J. Savenije, M.M. May, B. Dam, R. van de Krol, The origin of slow carrier transport in BiVO$_4$ thin film photoanodes: A time-resolved microwave conductivity study, J. Phys. Chem. Lett. 4 (2013) 2752–2757. https://doi.org/10.1021/jz4013257.

[64]   J. Wiktor, I. Reshetnyak, F. Ambrosio, A. Pasquarello, Comprehensive modeling of the band gap and absorption spectrum of $BiVO_4$, Phys. Rev. Mater. 1 (2017) 022401. https://doi.org/10.1103/PhysRevMaterials.1.022401.

[65]   J. Cen, S. Li, J. Zheng, F. Pan, Electron polarons in the subsurface layer of Mo/W-doped $BiVO_4$ surfaces, RSC Adv. 9 (2019) 819–823. https://doi.org/10.1039/C8RA09009B.

[66]   A. Chaudhuri, L. Mandal, X. Chi, M. Yang, M.C. Scott, M. Motapothula, X.J. Yu, P. Yang, Y. Shao-Horn, T. Venkatesan, A.T.S. Wee, A. Rusydi, Direct observation of anisotropic small-hole polarons in an orthorhombic structure of $BiVO_4$ films, Phys. Rev. B. 97 (2018) 195150. https://doi.org/10.1103/PhysRevB.97.195150.

[67]   B. Pattengale, J. Ludwig, J. Huang, Atomic insight into the W-doping effect on carrier dynamics and photoelectrochemical properties of $BiVO_4$ photoanodes, J. Phys. Chem. C. 120 (2016) 1421–1427. https://doi.org/10.1021/acs.jpcc.5b11451.

[68]   W. Luo, J. Wang, X. Zhao, Z. Zhao, Z. Li, Z. Zou, Formation energy and photoelectrochemical properties of $BiVO_4$ after doping at Bi3+ or V^{5+} sites with higher valence metal ions, Phys. Chem. Chem. Phys. 15 (2012) 1006–1013. https://doi.org/10.1039/C2CP43408C.

[69]   X. Zhao, J. Hu, S. Chen, Z. Chen, An investigation on the role of W doping in $BiVO_4$ photoanodes used for solar water splitting, Phys. Chem. Chem. Phys. 20 (2018) 13637–13645. https://doi.org/10.1039/C8CP01316K.

[70]   U. Prasad, J. Prakash, B. Azeredo, A. Kannan, Stoichiometric and non-stoichiometric tungsten doping effect in bismuth vanadate based photoactive material for photoelectrochemical water splitting, Electrochim. Acta 299 (2019) 262–272. https://doi.org/10.1016/j.electacta.2019.01.013.

[71]   M. Zhou, J. Bao, Y. Xu, J. Zhang, J. Xie, M. Guan, C. Wang, L. Wen, Y. Lei, Y. Xie, Photoelectrodes based upon Mo:$BiVO_4$ inverse opals for photoelectrochemical water splitting, ACS Nano 8 (2014) 7088–7098. https://doi.org/10.1021/nn501996a.

[72]   M. Favaro, R. Uecker, S. Nappini, I. Píš, E. Magnano, H. Bluhm, R. van de Krol, D.E. Starr, Chemical, structural, and electronic characterization of the (010) surface of single crystalline bismuth vanadate, J. Phys. Chem. C. (2018). (in press) https://doi.org/10.1021/acs.jpcc.8b09016.

[73]   L. Yang, Y. Xiong, W. Guo, J. Guo, D. Gao, Y. Zhang, P. Xiao, Mo^{6+} doped $BiVO_4$ with improved charge separation and oxidation kinetics for photoelectrochemical water splitting, Electrochim. Acta 256 (2017) 268–277. https://doi.org/10.1016/j.electacta.2017.09.186.

[74]   M. Tayebi, A. Tayyebi, B.-K. Lee, Improved photoelectrochemical performance of molybdenum (Mo)-doped monoclinic bismuth vanadate with increasing donor concentration, Catal. Today. (2018) (in press). https://doi.org/10.1016/j.cattod.2018.10.014.

[75]   M. Rohloff, B. Anke, S. Zhang, U. Gernert, C. Scheu, M. Lerch, A. Fischer, Mo-doped $BiVO_4$ thin films – high photoelectrochemical water splitting performance achieved by a tailored structure and morphology, Sustain. Energy Fuels 1 (2017) 1830–1846. https://doi.org/10.1039/C7SE00301C.

[76]   R.P. Antony, P.S. Bassi, F.F. Abdi, S.Y. Chiam, Y. Ren, J. Barber, J.S.C. Loo, L.H. Wong, Electrospun Mo-$BiVO_4$ for efficient photoelectrochemical water oxidation: Direct evidence of improved hole diffusion length and charge separation, Electrochimica Acta. 211 (2016) 173–182. https://doi.org/10.1016/j.electacta.2016.06.008.

[77]   A.J. Bard, Standard Potentials in Aqueous Solution, 1st edition, Taylor & Francis group, (1985), New York, https://doi.org/https://doi.org/10.1201/9780203738764.

[78]   R.P. Antony, T. Baikie, S.Y. Chiam, Y. Ren, R.R. Prabhakar, S.K. Batabyal, S.C.J. Loo, J. Barber, L.H. Wong, Catalytic effect of Bi^{5+} in enhanced solar water splitting of tetragonal $BiV_{0.8}Mo_{0.2}O_4$, Appl. Catal. Gen. 526 (2016) 21–27. https://doi.org/10.1016/j.apcata.2016.07.022.

[79]   O. Monfort, S. Sfaelou, L. Satrapinskyy, T. Plecenik, T. Roch, G. Plesch, P. Lianos, Comparative study between pristine and Nb-modified $BiVO_4$ films employed for photoelectrocatalytic production of H_2 by water splitting and for photocatalytic degradation of organic pollutants under simulated solar light, Catal. Today. 280 (2017) 51–57. https://doi.org/10.1016/j.cattod.2016.07.006.

[80]   Y. Zhang, Z. Yi, G. Wu, Q. Shen, Novel Y doped $BiVO_4$ thin film electrodes for enhanced photoelectric and photocatalytic performance, J. Photochem. Photobiol. Chem. 327 (2016) 25–32. https://doi.org/10.1016/j.jphotochem.2016.05.004.

[81]   B. Zhang, H. Zhang, Z. Wang, X. Zhang, X. Qin, Y. Dai, Y. Liu, P. Wang, Y. Li, B. Huang, Doping strategy to promote the charge separation in $BiVO_4$ photoanodes, Appl. Catal. B Environ. 211 (2017) 258–265. https://doi.org/10.1016/j.apcatb.2017.03.078.

[82]   C. Yin, S. Zhu, Z. Chen, W. Zhang, J. Gu, D. Zhang, One step fabrication of C-doped $BiVO_4$ with hierarchical structures for a high-performance photocatalyst under visible light irradiation, J. Mater. Chem. A. 1 (2013) 8367–8378. https://doi.org/10.1039/C3TA11833A.

[83]   X. Zhong, H. He, M. Yang, G. Ke, Z. Zhao, F. Dong, B. Wang, Y. Chen, X. Shi, Y. Zhou, In^{3+}-doped $BiVO_4$ photoanodes with passivated surface states for photoelectrochemical water oxidation, J. Mater. Chem. A 6 (2018) 10456–10465. https://doi.org/10.1039/C8TA01377B.

[84]   H. Ullah, A.A. Tahir, T.K. Mallick, Structural and electronic properties of oxygen defective and Se-doped p-type $BiVO_4(001)$ thin film for the applications of photocatalysis, Appl. Catal. B Environ. 224 (2018) 895–903. https://doi.org/10.1016/j.apcatb.2017.11.034.

[85]   T. Xia, M. Chen, L. Xiao, W. Fan, B. Mao, D. Xu, P. Guan, J. Zhu, W. Shi, Dip-coating synthesis of P-doped $BiVO_4$ photoanodes with enhanced photoelectrochemical performance, J. Taiwan Inst. Chem. Eng. 93 (2018) 582–589. https://doi.org/10.1016/j.jtice.2018.09.003.

[86]   L. Shan, G. Wang, J. Suriyaprakash, D. Li, L. Liu, L. Dong, Solar light driven pure water splitting of B-doped $BiVO_4$ synthesized via a sol–gel method, J. Alloys Compd. 636 (2015) 131–137. https://doi.org/10.1016/j.jallcom.2015.02.113.

[87]   B. Anke, M. Rohloff, M.G. Willinger, W. Hetaba, A. Fischer, M. Lerch, Improved photoelectrochemical performance of bismuth vanadate by partial O/F-substitution, Solid State Sci. 63 (2017) 1–8. https://doi.org/10.1016/j.solidstatesciences.2016.11.004.

[88]   A.C. Ulpe, B. Anke, S. Berendts, M. Lerch, T. Bredow, O/F-substitution in $BiVO_4$: Defect structures, phase stability and optical properties, Solid State Sci. 75 (2018) 39–44. https://doi.org/10.1016/j.solidstatesciences.2017.11.007.

[89]   G.V. Govindaraju, J.M. Morbec, G.A. Galli, K.-S. Choi, Experimental and computational investigation of lanthanide ion doping on $BiVO_4$ photoanodes for solar water splitting, J. Phys. Chem. C 122 (2018) 19416–19424. https://doi.org/10.1021/acs.jpcc.8b05503.

[90]   J. Quiñonero, T. Lana–Villarreal, R. Gómez, Improving the photoactivity of bismuth vanadate thin film photoanodes through doping and surface modification strategies, Appl. Catal. B Environ. 194 (2016) 141–149. https://doi.org/10.1016/j.apcatb.2016.04.057.

[91]   X. Zhao, J. Hu, B. Wu, A. Banerjee, S. Chakraborty, J. Feng, Z. Zhao, S. Chen, R. Ahuja, T.C. Sum, Z. Chen, Simultaneous enhancement in charge separation and onset potential for water oxidation in a $BiVO_4$ photoanode by W–Ti codoping, J. Mater. Chem. A. 6 (2018) 16965–16974. https://doi.org/10.1039/C8TA05491F.

[92]   H. He, S.P. Berglund, A.J.E. Rettie, W.D. Chemelewski, P. Xiao, Y. Zhang, C.B. Mullins, Synthesis of $BiVO_4$ nanoflake array films for photoelectrochemical water oxidation, J. Mater. Chem. A. 2 (2014) 9371–9379. https://doi.org/10.1039/C4TA00895B.

[93]   P.S. Bassi, L. Xianglin, Y. Fang, J.S.C. Loo, J. Barber, L.H. Wong, Understanding charge transport in non-doped pristine and surface passivated hematite (Fe_2O_3) nanorods under front and backside illumination in the context of light induced water splitting, Phys. Chem. Chem. Phys. 18 (2016) 30370–30378. https://doi.org/10.1039/C6CP05379C.

[94]   X. Chang, T. Wang, P. Zhang, Y. Wei, J. Zhao, J. Gong, Stable aqueous photoelectrochemical CO_2 reduction by a Cu_2O dark cathode with improved selectivity for carbonaceous products, Angew. Chem. Int. Ed. 55 (2016) 8840–8845. https://doi.org/10.1002/anie.201602973.

[95]   J. Gan, X. Lu, B.B. Rajeeva, R. Menz, Y. Tong, Y. Zheng, Efficient photoelectrochemical water oxidation over hydrogen-reduced nanoporous $BiVO_4$ with Ni–Bi electrocatalyst, ChemElectroChem 2 (2015) 1385–1395. https://doi.org/10.1002/celc.201500091.

[96]   S. Xiao, H. Chen, Z. Yang, X. Long, Z. Wang, Z. Zhu, Y. Qu, S. Yang, Origin of the Different photoelectrochemical performance of mesoporous $BiVO_4$ photoanodes between the $BiVO_4$ and the FTO side illumination, J. Phys. Chem. C 119 (2015) 23350–23357. https://doi.org/10.1021/acs.jpcc.5b07505.

[97]   B.J. Trześniewski, W.A. Smith, Photocharged BiVO4 photoanodes for improved solar water splitting, J. Mater. Chem. A. 4 (2016) 2919–2926. https://doi.org/10.1039/C5TA04716A.

[98]   B.J. Trześniewski, I.A. Digdaya, T. Nagaki, S. Ravishankar, I. Herraiz-Cardona, D.A. Vermaas, A. Longo, S. Gimenez, W.A. Smith, Near-complete suppression of surface losses and total internal quantum efficiency in $BiVO_4$ photoanodes, Energy Environ. Sci. 10 (2017) 1517–1529. https://doi.org/10.1039/C6EE03677E.

[99]   A.J. Bard, A.B. Bocarsly, F.R.F. Fan, E.G. Walton, M.S. Wrighton, The concept of Fermi level pinning at semiconductor/liquid junctions. Consequences for energy conversion efficiency and selection of useful solution redox couples in solar devices, J. Am. Chem. Soc. 102 (1980) 3671–3677. https://doi.org/10.1021/ja00531a001.

[100] S.R. Morrison, Electrochemistry at semiconductor and oxidized metal electrodes, Springer US, 1980.

[101] R. van de Krol, M. Grätzel, eds., Photoelectrochemical hydrogen production, Springer US, 2012.

[102] C. Du, X. Yang, M.T. Mayer, H. Hoyt, J. Xie, G. McMahon, G. Bischoping, D. Wang, Hematite-based water splitting with low turn-on voltages, Angew. Chem. Int. Ed. 52 (2013) 12692–12695. https://doi.org/10.1002/anie.201306263.

[103] Y.V. Pleskov, Solar energy conversion: A photoelectrochemical approach, Springer-Verlag, Berlin Heidelberg, 1990.

[104] S. Byun, B. Kim, S. Jeon, B. Shin, Effects of a SnO_2 hole blocking layer in a $BiVO_4$-based photoanode on photoelectrocatalytic water oxidation, J. Mater. Chem. A 5 (2017) 6905–6913. https://doi.org/10.1039/C7TA00806F.

[105] S. Byun, G. Jung, S.-Y. Moon, B. Kim, J.Y. Park, S. Jeon, S.-W. Nam, B. Shin, Compositional engineering of solution-processed $BiVO_4$ photoanodes toward highly efficient photoelectrochemical water oxidation, Nano Energy 43 (2018) 244–252. https://doi.org/10.1016/j.nanoen.2017.11.034.

[106] B. Lamm, L. Zhou, P. Rao, M. Stefik, Atomic layer deposition of space-efficient SnO_2 underlayers for bivo$_4$ host–guest architectures for photoassisted water splitting, ChemSusChem. (2019) in press. https://doi.org/10.1002/cssc.201802566.

[107] W. Zhang, D. Yan, X. Tong, M. Liu, Ultrathin lutetium oxide film as an epitaxial hole-blocking layer for crystalline bismuth vanadate water splitting photoanodes, Adv. Funct. Mater. 28 (2018) 1705512. https://doi.org/10.1002/adfm.201705512.

[108] S.J. Hong, S. Lee, J.S. Jang, J.S. Lee, Heterojunction $BiVO_4$/WO_3 electrodes for enhanced photoactivity of water oxidation, Energy Environ. Sci. 4 (2011) 1781–1787. https://doi.org/10.1039/C0EE00743A.

[109] Y. Liang, T. Tsubota, L.P.A. Mooij, R. van de Krol, Highly improved quantum efficiencies for thin film $BiVO_4$ photoanodes, J. Phys. Chem. C. 115 (2011) 17594–17598. https://doi.org/10.1021/jp203004v.

[110] P. Chatchai, Y. Murakami, S.–. Kishioka, A.Y. Nosaka, Y. Nosaka, FTO/SnO_2/$BiVO_4$ composite photoelectrode for water oxidation under visible light irradiation, Electrochem. Solid-State Lett. 11 (2008) H160–H163. https://doi.org/10.1149/1.2906109.

[111] M.G. Lee, D.H. Kim, W. Sohn, C.W. Moon, H. Park, S. Lee, H.W. Jang, Conformally coated $BiVO_4$ nanodots on porosity-controlled WO_3 nanorods as highly efficient type II heterojunction photoanodes for water oxidation, Nano Energy 28 (2016) 250–260. https://doi.org/10.1016/j.nanoen.2016.08.046.

[112] K.S. Joya, Y.F. Joya, K. Ocakoglu, R. van de Krol, Water-splitting catalysis and solar fuel devices: artificial leaves on the move, Angew. Chem. Int. Ed. 52 (2013) 10426–10437. https://doi.org/10.1002/anie.201300136.

[113] K. Sivula, Metal oxide photoelectrodes for solar fuel production, surface traps, and catalysis, J. Phys. Chem. Lett. 4 (2013) 1624–1633. https://doi.org/10.1021/jz4002983.

[114] M. Zhong, T. Hisatomi, T. Minegishi, H. Nishiyama, M. Katayama, T. Yamada, K. Domen, Bulky crystalline $BiVO_4$ thin films for efficient solar water splitting, J. Mater. Chem. A 4 (2016) 9858–9864. https://doi.org/10.1039/C6TA03072F.

[115] L. Zhang, C.Y. Lin, V.K. Valev, E. Reisner, U. Steiner, J.J. Baumberg, Plasmonic enhancement in $BiVO_4$ photonic crystals for efficient water splitting, Small 10 (2014) 3970–3978. https://doi.org/10.1002/smll.201400970.

[116] C. Ding, J. Shi, D. Wang, Z. Wang, N. Wang, G. Liu, F. Xiong, C. Li, Visible light driven overall water splitting using cocatalyst/$BiVO_4$ photoanode with minimized bias, Phys. Chem. Chem. Phys. 15 (2013) 4589–4595. https://doi.org/10.1039/C3CP50295C.

[117] T.W. Kim, K.-S. Choi, Nanoporous $BiVO_4$ photoanodes with dual-layer oxygen evolution catalysts for solar water splitting, Science 343 (2014) 990–994. https://doi.org/10.1126/science.1246913.

[118] J.H. Kim, J.W. Jang, H.J. Kang, G. Magesh, J.Y. Kim, J.H. Kim, J. Lee, J.S. Lee, Palladium oxide as a novel oxygen evolution catalyst on $BiVO_4$ photoanode for photoelectrochemical water splitting, J. Catal. 317 (2014) 126–134. https://doi.org/10.1016/j.jcat.2014.06.015.

[119] Y. Liang, J. Messinger, Improving $BiVO_4$ photoanodes for solar water splitting through surface passivation, Phys. Chem. Chem. Phys. 16 (2014) 12014–12020. https://doi.org/10.1039/C4CP00674G.

[120] L. Chen, F.M. Toma, J.K. Cooper, A. Lyon, Y. Lin, I.D. Sharp, J.W. Ager, Mo-Doped $BiVO_4$ photoanodes synthesized by reactive sputtering, ChemSusChem 8 (2015) 1066–1071. https://doi.org/10.1002/cssc.201402984.

[121] M. de Respinis, K.S. Joya, H.J.M. De Groot, F. D'Souza, W.A. Smith, R. van de Krol, B. Dam, Solar Water Splitting Combining a $BiVO_4$ Light Absorber with a Ru-Based Molecular Cocatalyst, J. Phys. Chem. C. 119 (2015) 7275–7281. https://doi.org/10.1021/acs.jpcc.5b00287.

[122] M.G. Lee, C.W. Moon, H. Park, W. Sohn, S.B. Kang, S. Lee, K.J. Choi, H.W. Jang, Dominance of plasmonic resonant energy transfer over direct electron transfer in

substantially enhanced water oxidation activity of $BiVO_4$ by Shape-Controlled Au Nanoparticles, Small. 13 (2017) 1701644. https://doi.org/10.1002/smll.201701644.

[123] K.-H. Ye, Z. Wang, J. Gu, S. Xiao, Y. Yuan, Y. Zhu, Y. Zhang, W. Mai, S. Yang, Carbon quantum dots as a visible light sensitizer to significantly increase the solar water splitting performance of bismuth vanadate photoanodes, Energy Environ. Sci. 10 (2017) 772–779. https://doi.org/10.1039/C6EE03442J.

[124] M.-W. Kim, K. Kim, T.Y. Ohm, B. Joshi, E. Samuel, M.T. Swihart, H. Yoon, H. Park, S.S. Yoon, Mo-doped $BiVO_4$ nanotextured pillars as efficient photoanodes for solar water splitting, J. Alloys Compd. 726 (2017) 1138–1146. https://doi.org/10.1016/j.jallcom.2017.07.260.

[125] X. Zhang, R. Wang, F. Li, Z. An, M. Pu, X. Xiang, Enhancing photoelectrochemical water oxidation efficiency of $bivo_4$ photoanodes by a hybrid structure of layered double hydroxide and graphene, Ind. Eng. Chem. Res. 56 (2017) 10711–10719. https://doi.org/10.1021/acs.iecr.7b02960.

[126] F.S. Hegner, I. Herraiz-Cardona, D. Cardenas-Morcoso, N. López, J.-R. Galán-Mascarós, S. Gimenez, Cobalt hexacyanoferrate on $BiVO_4$ photoanodes for robust water splitting, ACS Appl. Mater. Interfaces. 9 (2017) 37671–37681. https://doi.org/10.1021/acsami.7b09449.

[127] Y. Wang, F. Li, X. Zhou, F. Yu, J. Du, L. Bai, L. Sun, Highly efficient photoelectrochemical water splitting with an immobilized molecular Co_4O_4 cubane catalyst, Angew. Chem. Int. Ed. 56 (2017) 6911–6915. https://doi.org/10.1002/anie.201703039.

[128] M. Chhetri, S. Dey, C.N.R. Rao, Photoelectrochemical oxygen evolution reaction activity of amorphous Co–La double hydroxide-$BiVO_4$ fabricated by pulse plating electrodeposition, ACS Energy Lett. 2 (2017) 1062–1069. https://doi.org/10.1021/acsenergylett.7b00247.

[129] J.K. Kim, X. Shi, M.J. Jeong, J. Park, H.S. Han, S.H. Kim, Y. Guo, T.F. Heinz, S. Fan, C.-L. Lee, J.H. Park, X. Zheng, Enhancing Mo:$BiVO_4$ solar water splitting with patterned Au nanospheres by plasmon-induced energy transfer, Adv. Energy Mater. 8 (2018) 1701765. https://doi.org/10.1002/aenm.201701765.

[130] M.-W. Kim, E. Samuel, K. Kim, H. Yoon, B. Joshi, M.T. Swihart, S.S. Yoon, Tuning the morphology of electrosprayed $BiVO_4$ from nanopillars to nanoferns via pH control for solar water splitting, J. Alloys Compd. 769 (2018) 193–200. https://doi.org/10.1016/j.jallcom.2018.07.167.

[131] P. Guan, H. Bai, F. Wang, H. Yu, D. Xu, B. Chen, T. Xia, W. Fan, W. Shi, Boosting water splitting performance of bivo₄ photoanode through selective surface decoration of Ag₂S, ChemCatChem. 10 (2018) 4927–4933. https://doi.org/10.1002/cctc.201801199.

[132] X. Han, Y. Wei, J. Su, Y. Zhao, Low-cost oriented hierarchical growth of BiVO₄/rGO/NiFe nanoarrays photoanode for photoelectrochemical water splitting, ACS Sustain. Chem. Eng. 6 (2018) 14695–14703. https://doi.org/10.1021/acssuschemeng.8b03259.

[133] M. Zhang, R.P. Antony, S.Y. Chiam, F.F. Abdi, L.H. Wong, Understanding the roles of NiOₓ in enhancing the photoelectrochemical performance of BiVO₄ photoanodes for solar water splitting, ChemSusChem. (2019) in press. https://doi.org/10.1002/cssc.201801780.

[134] B.-R. Wulan, S.S. Yi, S.J. Li, Y.X. Duan, J.-M. Yan, X.B. Zhang, Q. Jiang, Non-noble-metal bismuth nanoparticle-decorated bismuth vanadate nanoarray photoanode for efficient water splitting, Mater. Chem. Front. 2 (2018) 1799–1804. https://doi.org/10.1039/C8QM00239H.

[135] L. Li, J. Li, J. Bai, Q. Zeng, L. Xia, Y. Zhang, S. Chen, Q. Xu, B. Zhou, Serial hole transfer layers for a BiVO₄ photoanode with enhanced photoelectrochemical water splitting, Nanoscale 10 (2018) 18378–18386. https://doi.org/10.1039/C8NR06342G.

[136] M.-W. Kim, K. Kim, T.Y. Ohm, H. Yoon, B. Joshi, E. Samuel, M.T. Swihart, S.K. Choi, H. Park, S.S. Yoon, Electrosprayed BiVO₄ nanopillars coated with atomic-layer-deposited ZnO/TiO₂ as highly efficient photoanodes for solar water splitting, Chem. Eng. J. 333 (2018) 721–729. https://doi.org/10.1016/j.cej.2017.09.130.

[137] M.N. Shaddad, D. Cardenas-Morcoso, P. Arunachalam, M. García-Tecedor, M.A. Ghanem, J. Bisquert, A. Al-Mayouf, S. Gimenez, Enhancing the optical absorption and interfacial properties of bivo₄ with ag₃po₄ nanoparticles for efficient water splitting, J. Phys. Chem. C 122 (2018) 11608–11615. https://doi.org/10.1021/acs.jpcc.8b00738.

[138] J. Wei, C. Zhou, Y. Xin, X. Li, L. Zhao, Z. Liu, Cooperation effect of heterojunction and co-catalyst in BiVO₄/Bi₂S₃/NiOOH photoanode for improving photoelectrochemical performances, New J. Chem. 42 (2018) 19415–19422. https://doi.org/10.1039/C8NJ04846K.

[139] H. Saada, R. Abdallah, B. Fabre, D. Floner, S. Fryars, A. Vacher, V. Dorcet, C. Meriadec, S. Ababou-Girard, G. Loget, Boosting the performance of BiVO₄ prepared

through alkaline electrodeposition with an amorphous Fe Co-Catalyst, ChemElectroChem 6 (2019) 613–617. https://doi.org/10.1002/celc.201801443.

[140] S. Wang, T. He, J.-H. Yun, Y. Hu, M. Xiao, A. Du, L. Wang, New iron-cobalt oxide catalysts promoting BiVO$_4$ films for photoelectrochemical water splitting, Adv. Funct. Mater. 28 (2018) 1802685. https://doi.org/10.1002/adfm.201802685.

[141] M. Kan, D. Xue, A. Jia, X. Qian, D. Yue, J. Jia, Y. Zhao, A highly efficient nanoporous BiVO$_4$ photoelectrode with enhanced interface charge transfer Co-catalyzed by molecular catalyst, Appl. Catal. B Environ. 225 (2018) 504–511. https://doi.org/10.1016/j.apcatb.2017.12.014.

[142] X. Lv, X. Xiao, M. Cao, Y. Bu, C. Wang, M. Wang, Y. Shen, Efficient carbon dots/NiFe-layered double hydroxide/BiVO$_4$ photoanodes for photoelectrochemical water splitting, Appl. Surf. Sci. 439 (2018) 1065–1071. https://doi.org/10.1016/j.apsusc.2017.12.182.

[143] Y. Gao, Y. Li, G. Yang, S. Li, N. Xiao, B. Xu, S. Liu, P. Qiu, S. Hao, L. Ge, Fe$_2$TiO$_5$ as an efficient co-catalyst to improve the photoelectrochemical water splitting performance of BiVO$_4$, ACS Appl. Mater. Interfaces. 10 (2018) 39713–39722. https://doi.org/10.1021/acsami.8b14141.

[144] L. Wang, J. Su, L. Guo, Hierarchical growth of a novel Mn-Bi coupled BiVO$_4$ arrays for enhanced photoelectrochemical water splitting, Nano Res. 12 (2019) 575-580. https://doi.org/10.1007/s12274-018-2256-9.

[145] Q. Wang, T. Niu, L. Wang, C. Yan, J. Huang, J. He, H. She, B. Su, Y. Bi, FeF$_2$/BiVO$_4$ heterojuction photoelectrodes and evaluation of its photoelectrochemical performance for water splitting, Chem. Eng. J. 337 (2018) 506–514. https://doi.org/10.1016/j.cej.2017.12.126.

[146] J.K. Kim, Y. Cho, M.J. Jeong, B. Levy-Wendt, D. Shin, Y. Yi, D.H. Wang, X. Zheng, J.H. Park, Rapid formation of a disordered layer on monoclinic BiVO$_4$: Co-catalyst-free photoelectrochemical solar water splitting, ChemSusChem. 11 (2018) 933–940. https://doi.org/10.1002/cssc.201702173.

[147] W. Zhang, R. Li, X. Zhao, Z. Chen, A.W.-K. Law, K. Zhou, A cobalt-based metal–organic framework as cocatalyst on BiVO$_4$ photoanode for enhanced photoelectrochemical water oxidation, ChemSusChem. 11 (2018) 2710–2716. https://doi.org/10.1002/cssc.201801162.

[148] Y. Liu, Y. Jiang, F. Li, F. Yu, W. Jiang, L. Xia, Molecular cobalt salophen catalyst-integrated BiVO$_4$ as stable and robust photoanodes for photoelectrochemical

water splitting, J. Mater. Chem. A 6 (2018) 10761–10768.
https://doi.org/10.1039/C8TA01304G.

[149] Y. Hu, Y. Wu, J. Feng, H. Huang, C. Zhang, Q. Qian, T. Fang, J. Xu, P. Wang, Z. Li, Z. Zou, Rational design of electrocatalysts for simultaneously promoting bulk charge separation and surface charge transfer in solar water splitting photoelectrodes, J. Mater. Chem. A. 6 (2018) 2568–2576. https://doi.org/10.1039/C7TA10361A.

[150] Q. Pan, C. Zhang, Y. Xiong, Q. Mi, D. Li, L. Zou, Q. Huang, Z. Zou, H. Yang, Boosting charge separation and transfer by plasmon-enhanced $MoS_2/BiVO_4$ p–n heterojunction composite for efficient photoelectrochemical water splitting, ACS Sustain. Chem. Eng. 6 (2018) 6378–6387.
https://doi.org/10.1021/acssuschemeng.8b00170.

[151] F. Tang, W. Cheng, H. Su, X. Zhao, Q. Liu, Smoothing surface trapping states in 3D coral-like $CoOOH$-wrapped-$BiVO_4$ for efficient photoelectrochemical water oxidation, ACS Appl. Mater. Interfaces. 10 (2018) 6228–6234.
https://doi.org/10.1021/acsami.7b15674.

[152] X. Wang, K.-H. Ye, X. Yu, J. Zhu, Y. Zhu, Y. Zhang, Polyaniline as a new type of hole-transporting material to significantly increase the solar water splitting performance of $BiVO_4$ photoanodes, J. Power Sources 391 (2018) 34–40.
https://doi.org/10.1016/j.jpowsour.2018.04.074.

[153] L. Gao, F. Li, H. Hu, X. Long, N. Xu, Y. Hu, S. Wei, C. Wang, J. Ma, J. Jin, Dual modification of a $BiVO_4$ photoanode for enhanced photoelectrochemical performance, ChemSusChem. 11 (2018) 2502–2509. https://doi.org/10.1002/cssc.201800999.

[154] P. Subramanyam, T. Vinodkumar, D. Nepak, M. Deepa, C. Subrahmanyam, Mo-doped $BiVO_4$@reduced graphene oxide composite as an efficient photoanode for photoelectrochemical water splitting, Catal. Today. 325 (2019) 73–80.
https://doi.org/10.1016/j.cattod.2018.07.006.

[155] M. Wang, Q. Wang, P. Guo, Z. Jiao, In situ fabrication of nanoporous $BiVO_4/Bi_2S_3$ nanosheets for enhanced photoelectrochemical water splitting, J. Colloid Interface Sci. 534 (2019) 338–342. https://doi.org/10.1016/j.jcis.2018.09.056.

[156] D. Xue, M. Kan, X. Qian, Y. Zhao, A tandem water splitting cell based on nanoporous $BiVO_4$ photoanode cocatalyzed by ultrasmall cobalt borate sandwiched with conformal TiO_2 layers, ACS Sustain. Chem. Eng. 6 (2018) 16228–16234.
https://doi.org/10.1021/acssuschemeng.8b03078.

[157] T.-G. Vo, Y. Tai, C.Y. Chiang, Multifunctional ternary hydrotalcite-like nanosheet arrays as an efficient co-catalyst for vastly improved water splitting performance on bismuth vanadate photoanode, J. Catal. 370 (2019) 1–10. https://doi.org/10.1016/j.jcat.2018.12.002.

[158] P. Guan, H. Bai, F. Wang, H. Yu, D. Xu, W. Fan, W. Shi, In-situ anchoring Ag through organic polymer for configuring efficient plasmonic BiVO$_4$ photoanode, Chem. Eng. J. 358 (2019) 658–665. https://doi.org/10.1016/j.cej.2018.10.066.

[159] M.N. Shaddad, P. Arunachalam, J. Labis, M. Hezam, A.M. Al-Mayouf, Fabrication of robust nanostructured (Zr)BiVO$_4$/nickel hexacyanoferrate core/shell photoanodes for solar water splitting, Appl. Catal. B Environ. 244 (2019) 863–870. https://doi.org/10.1016/j.apcatb.2018.11.079.

[160] A. Kafizas, X. Xing, S. Selim, C.A. Mesa, Y. Ma, C. Burgess, M.A. McLachlan, J.R. Durrant, Ultra-thin Al$_2$O$_3$ coatings on BiVO$_4$ photoanodes: Impact on performance and charge carrier dynamics, Catal. Today. 321–322 (2019) 59–66. https://doi.org/10.1016/j.cattod.2017.11.014.

[161] T.-G. Vo, Y. Tai, C.-Y. Chiang, Novel hierarchical ferric phosphate/bismuth vanadate nanocactus for highly efficient and stable solar water splitting, Appl. Catal. B Environ. 243 (2019) 657–666. https://doi.org/10.1016/j.apcatb.2018.11.001.

[162] S.Y. Jeong, H.-M. Shin, Y.-R. Jo, Y.J. Kim, S. Kim, W.-J. Lee, G.J. Lee, J. Song, B.J. Moon, S. Seo, H. An, S.H. Lee, Y.M. Song, B.-J. Kim, M.-H. Yoon, S. Lee, Plasmonic silver nanoparticle-impregnated nanocomposite BiVO$_4$ photoanode for plasmon-enhanced photocatalytic water splitting, J. Phys. Chem. C. 122 (2018) 7088–7093. https://doi.org/10.1021/acs.jpcc.8b00220.

[163] L. Zhang, L.O. Herrmann, J.J. Baumberg, Size dependent plasmonic effect on BiVO$_4$ photoanodes for solar water splitting, Sci. Rep. 5 (2015) 16660. https://doi.org/10.1038/srep16660.

[164] C. Zachäus, F.F. Abdi, L.M. Peter, R. van de Krol, Photocurrent of BiVO$_4$ is limited by surface recombination, not surface catalysis, Chem. Sci. 8 (2017) 3712–3719. https://doi.org/10.1039/C7SC00363C.

[165] X. Zhang, Y.L. Chen, R.-S. Liu, D.P. Tsai, Plasmonic photocatalysis, Rep. Prog. Phys. 76 (2013) 046401. https://doi.org/10.1088/0034-4885/76/4/046401.

[166] S. Linic, P. Christopher, D.B. Ingram, Plasmonic-metal nanostructures for efficient conversion of solar to chemical energy, Nat. Mater. 10 (2011) 911–921. https://doi.org/10.1038/nmat3151.

Photoelectrochemical Water Splitting: Materials and Applications   Materials Research Forum LLC
Materials Research Foundations **70** (2020) 183-213   https://doi.org/10.21741/9781644900734-7

Chapter 7

Noble Materials for Photoelectrochemical Water Splitting

Bijoy Tudu, Rajashree Bortamuly, Pranjal Saikia*

Department of Applied Sciences (Chemical Science Division), Gauhati University, Guwahati-781014, Assam, India

*psjorhat@gmail.com, pranjalsaikia@gauhati.ac.in

Abstract

Development of stable and highly efficient photoelectrodes capable of harvesting solar energy is an essential need of the hour to deal with the global energy demand. Photoelectrochemical (PEC) water splitting is a clean methodology utilizing solar light to produce clean and sustainable hydrogen energy. The best approach to achieve high solar-to-hydrogen energy conversion efficiency is the design of efficient noble materials capable of plasmonic absorption, competent charge separation and their utilization. This chapter presents the recent development on noble metals based photocatalysts that have been widely applied in PEC water splitting.

Keywords

Water Splitting, H_2 Evolution, Noble Metals, Photoelectrode, Photocurrent Density, Plasmonic Absorption, Schottky Junction

Contents

1.   Introduction

The alarming rate of the world population growth and technology dependent lifestyle has resulted in the need of large amounts of energy. The global energy consumption is rapidly increasing and expected to at least double by 2050. Therefore, we have to be able to find some renewable and environmentally clean energy resources in quickly [1,2]. PEC water splitting is a clean approach for solar energy conversion efficiently into hydrogen energy [3]. Since, hydrogen has already been recognized as the future fuel for its storable potentially and clean nature, it can be generated from natural water through the process of photocatalytic water splitting.

Water splitting into hydrogen and oxygen with minimum possible consumption of energy is an important chemical process which resembles the phenomenon of natural photosynthesis. Since Fujishima and Honda first disclosed the photoelectrochemical decomposition of water to generate hydrogen (H$_2$) by illuminating TiO$_2$ electrode in a photoelectrochemical cell, this has been gaining interest from scientists of all corners of the globe [4]. They used Pt and TiO$_2$ electrodes as cathode and anode respectively and hydrogen (H$_2$) and oxygen (O$_2$) were found to be evolved at respective cathode and anode during the reaction. Later on, Bard established the idea of photocatalysis to generate hydrogen. These two breakthroughs were regarded as pioneering works for effective utilization of the solar spectrum to decompose water to get hydrogen [5-7]. In PEC water splitting, photon energy is absorbed by a semiconductor or photoelectrode to generate holes (h$^+$) and electrons (e$^-$) at valence band (VB) and conduction band (CB) respectively, which are also known as charge carriers. The movement of these charge carriers leads to the formation of an intermediate, which is eventually responsible for the water splitting reaction, by decreasing the activation energy [8]. The role of the semiconductor in PEC reaction is crucial because the fate of the reaction depends on several properties of the semiconductor such as stability, band edge and the charge carrying ability etc. [9,10]. For complete water splitting, the positions of CB and VB of the semiconductor are important which should be at least 1.23 eV. If the conduction band edge is more negative than the reduction potential of H$^+$ to H$_2$, hydrogen evolution is favoured. On the other hand, if the valence band edge is more positive than the oxidation potential of water, it leads to oxygen evolution. In a PEC cell, working electrode is the semiconductor photoelectrode and Pt acts as the counter electrode. A reference electrode (saturated Ag/AgCl) having well-known electrode potential and electrolytes between two electrodes complete the setup [7,11]. The water splitting requires relatively high energy to occur, more than the energy released during water formation. Therefore, an artificial way to promote the water splitting reaction (H$_2$O→O$_2$+H$_2$) is being assisted by semiconductor materials or catalysts [12]. Different metal oxides have been widely used as stable and efficient photoelectrodes to improve photoelectrochemical (PEC) performance [13-17]. Among these, TiO$_2$ has been the most widely used as photoanode in PEC water splitting cells because it is inexpensive, non-toxic, highly stable and an efficient photocatalyst. However, it has also certain drawbacks like less surface area, inefficiency in utilizing entire solar spectrum and lesser charge carriers' separation ability [18]. Therefore, some modifications needed to be developed to improve its photocatalytic performance. Inclusions of photosensitizers or co-catalysts were attempted to improve hydrogen production activity, where the hole scavengers played a crucial role. Examples of some common hole scavengers include EDTA, methanol, ethanol, and lactic acid [19]. The noble metals served as active co-catalysts in semiconductor based catalysts applied to photocatalytic or photoelectrochemical water splitting. The surface modifications of semiconductors with noble metals to form metal-semiconductor systems were very crucial towards achieving high PEC performance. Here, active sites as well as trapping sites were made available by the metals for the generated charges thereby helping charge separation. As a consequence, local surface plasmon resonance (LSPR) came into effect for efficient utilization of plasmatic carriers. Depending on the direction of charge transfer, two catalytic mechanisms, namely, the Schottky junction and the LSPR effect were known in metal-semiconductor photocatalysts [8,20]. The larger work function of noble metals leading to creation of Schottky barrier efficiently separate the electron-hole pair contributing towards enhanced PEC performance [21]. One interesting behavior of some noble metal nanoparticles is that they can absorb visible light by virtue of LSPR and simultaneously can make some semiconductors visible light active by contraction of their band gap [22-24]. There are reports on design of photocatalysts which are efficient for hydrogen generation from water using a sacrificial agent. Metallic NPs, particularly these of silver, gold and platinum or a combination of these metals (Ag-Pt, Au-Pt, and Au-Ag) and various oxides are used to obtain shape and size selective nanostructures that give varied optical and surface properties [25,26]. The attractive options widely known are formation of alloy or bimetals of platinum with gold which have been proven to progress the catalytic activity for hydrogen generation using solar light and degradation of organic dyes or phenol [27]. Noted improvement in photocatalytic activity of TiO_2 was achieved by modifying it with bimetals of platinum-palladium in addition to platinum-gold [28,29]. Also, palladium-gold, platinum-copper, platinum-nickel, platinum-tin, platinum-iron, palladium-copper, gold-silver and copper-silver are receiving much interest [30-34]. Furthermore, with the idea of improving PEC performance of the electro-catalysts, loadings of metal NPs on various supports such as g-C_3N_4 and graphene based materials have become prominent in the field of photocatalysis. Interestingly, g-C_3N_4 itself is very efficient for PEC water splitting due to outstanding light absorption capability in visible region and high stability wherein deposition of noble metals such as Au, Pt, etc. further improve PEC performance through inhibition of charge recombination and acceleration of charge transportation. Recently, graphene has substantially been studied for water splitting due to its characteristic property of exceptional electronic mobility. Moreover, graphene is a very high surface area material due to the presence of exclusive surface atoms and a π-conjugated configuration. These properties help graphene to become a better material for photocatalytic H_2 evolution by lowering the rate of charge recombination, enhancing interfacial charge movements and increasing effective adsorption sites. The reduced graphene oxide (rGO) can act as an electron sink. Therefore, it can separate the generated electron-hole pairs easily. Interestingly, the functional groups containing oxygen on the GO are in competition with the H^+ ions to take those electrons. Thus, it is imperative to prepare graphene or GO-based materials for this purpose [8,35-39]. Even though, noble materials exhibit efficient photo electrochemical (PEC) activities, their practical utility is hindered by high cost and lesser availability. In this regard, non-noble metals have been considered most suitable alternatives for their cost effectiveness, appropriate work function and superb conducting nature. However, no systematic report on composites of noble or non-noble metals for PEC water splitting enlightening the function of these metals has been available in literature.

In this chapter, we aimed to sum up the advances regarding utility of noble metals for PEC water splitting to divulge the relation of metals and metal hybridized semiconductor systems for their effectiveness. Also, we intend to discuss monometallic and bimetallic based noble materials systematically through focusing on their characteristic properties as well as utilization of these metals based composites for the efficient photoelectrodes in PEC water splitting.

2.   Fundamental properties of noble metals for photocatalytic activity

2.1   Fundamentals of the Localized Surface Plasmon Resonance (LSPR)

Any material with nanoshape and considerable free electrons or holes can work as plasmonic media. Plasmon resonance leads to augmentation of the nanoparticle's local electric field. When light (with a wavelength far greater than the NP size) is illuminated on a small spherical metallic NP, an unsymmetrical electron distribution occurs. This unsymmetrical charge distribution results in an electric field in and out of the metal NP with a direction opposite to that of the electric field of the light (Fig. 1). This leads to localization of electron density and thereby the Coulombic restoring force is generated. The resonance between the oscillations and the incident light is known as the localized surface plasmon resonance (LSPR). Hence the LSPR property bestows the metal nanocrystals with large absorption and scattering cross-sections and intensifies local electromagnetic field in the near-field region of the plasmonic metal nanocrystals surface [40].

2.2   Schottky junction

The potential energy junction of electrons in metal-semiconductor interface is called Schottky junction. In case of n-type semiconductor, the work function of metal is relatively higher than that of the semiconductor, thus the transfer of free electrons from the semiconductor to the metal prominently occurs until the Fermi levels of both metal and semiconductor matches. Consequently, the formation of upward band bending occurs by the charge diffusion process and the junction that is formed is known as Schottky junction. During light illumination, the rapid trapping of photogenerated electrons by metal from the conduction band (CB) of semiconductor occurs, leading to high separation and transportation of electron-hole pairs, the metal acting as co-catalyst [41].

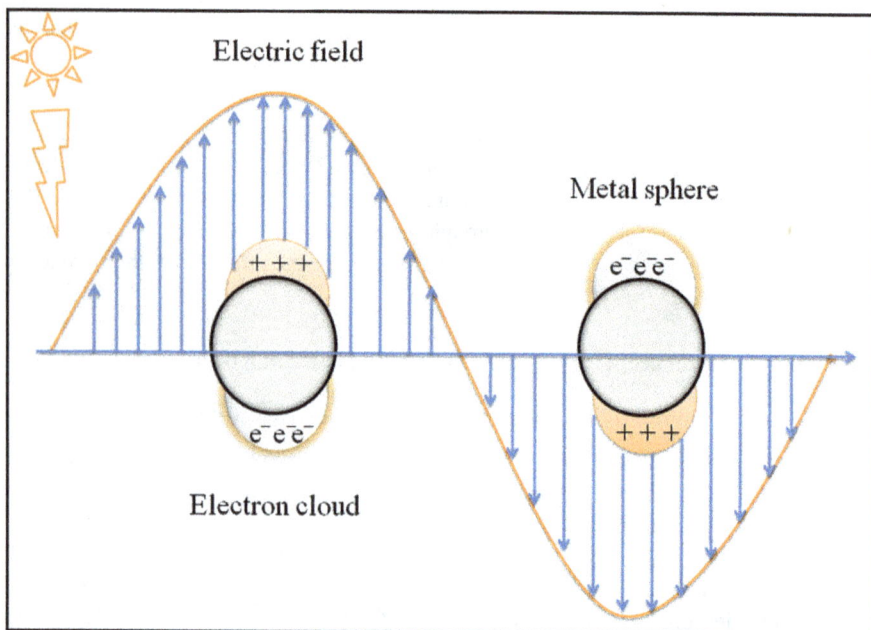

Figure 1: LSPR of noble metal nanoparticles, a collective electrons density oscillation caused by the electric field component of incident solar light.

3.   Photoelectrodes materials

Since 1972, many semiconductors have been engineered for photoelectrodes for use in photoelectrochemical (PEC) cells. The materials for photoelectrodes preparation must have the properties of light absorption, charge separation, migration and ability to transfer to the electrolyte solution for redox reactions. Some of the photoelectrode materials used in PEC water splitting are discussed as follows:

3.1   Titania (TiO₂)

Titania, being the first ever used material for photoelectrochemical water splitting reaction, has retained its glory as photocatalyst or photoelectrode material till date. It has a wide band gap with 3.2 eV in case of anatase and 3.0 eV for rutile phase. TiO_2 is a better candidate for use as photoelectrode due to its abundance, non-toxicity and photochemical stability under experimental conditions. But due to its large band gap, it is able to absorb only the ultraviolet (UV) light (< 5% of total solar radiation) which leads

Photoelectrochemical Water Splitting: Materials and Applications   Materials Research Forum LLC
Materials Research Foundations 70 (2020) 183-213   https://doi.org/10.21741/9781644900734-7

to an extremely low theoretical solar to hydrogen (STH) conversion efficiency (η=1.3% in case of anatase and η=2.2% in case of rutile TiO_2). For the last few decades, many efforts have been made to dope TiO_2 with metals to enhance visible-light induced photocatalytic activity. Doping enhances its light absorption spectral range from UV to the visible region [5,42].

3.2 Haematite (Fe₂O₃)

Fe_2O_3 has been recognized as a most favorable material for water splitting reactions. It has band gap of 2.1 eV, and hence can absorb considerable amount of solar light. In comparison to other photoanode-materials, Fe_2O_3 with enormous stability and compatibility is easily available at low cost. The preparation routes, surface-morphology of Fe_2O_3 as well as doping with metal co-catalysts are important in controlling the efficiency of photoelectrochemical (PEC) water splitting [43,44].

3.3 Zinc oxide (ZnO)

ZnO is of great utilization in water splitting. It has a wide energy band gap of 3.37 eV and therefore optoelectronically comparable to TiO_2. Owing to its faint activity towards visible light and low quantum efficiency, the photocatalytic efficiency of ZnO is comparatively low. Surface modification by doping metals and tailoring their nanostructures through deposition techniques and annealing have been helpful to improve its activity [45].

4. Fundamental role of noble materials in PEC water splitting

The past attempts for the designing of semiconductor based photoelectrodes; concept and morphology modification techniques for improving photoelectrochemical water splitting to produce hydrogen (H_2) are briefly presented. The crucial roles of noble metals as co-catalysts loaded onto photoelectrode for improving PEC activities are highlighted in the following paragraphs. Fig. 2 represents the possible photocatalytic mechanism for improving PEC water splitting. In this mechanism, the noble metals serve as active sites to absorb the plasmonic light due to LSPR effect producing hot electrons. These electrons are transferred to conduction band (CB) of the semiconductor where the electrons are involved in reduction reaction to produce hydrogen. Also, there is a possibility of rapid transfer of these electrons to the Schottky junction established in the metal-semiconductor interface. This process inhibits the electron-hole recombination and hence increasing the photocatalytic evolution of H_2. To split the water, the band gap of semiconductor must be greater or equal to 1.23 eV which is a reduction potential of water. In a photoelectrochemical cell, the evolution of hydrogen is measured in terms of photocurrent density. High values of photocurrent generation indicate more evolutions of hydrogen. The photocurrent densities of some monometallic and bimetallic noble materials are listed in Tables 1 and 2 respectively.

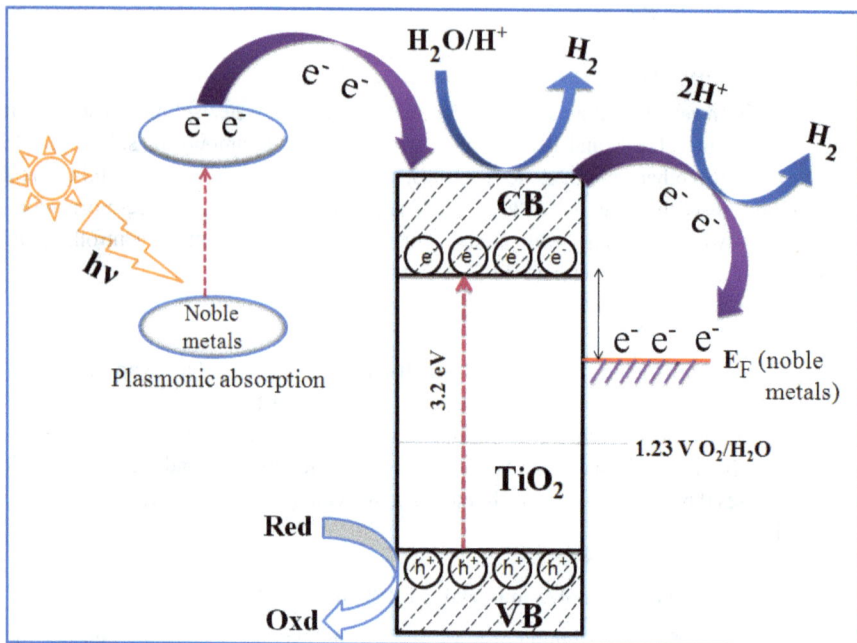

Figure 2: Possible mechanism of electron transfer during photocatalysis for H_2 evolution from pure water. Visible light absorption occurs due to plasmonic noble metals.

Table 1. Photocurrent measurements of single noble metals based photoanode in PEC water splitting

Photoelctrode	Metal loading method	Photocurrent density	Electrolyte	Ref.
Pt/TiO$_2$	Hydrothermal deposition	0.67 mAcm^{-2} at 1.23V$_{RHE}$	0.1 M NaOH	51
Pt/TiO$_2$(TNTF)	Hydrothermal and sputtering	3.8 mAcm^{-2} at 1.0V$_{Ag/AgCl}$	0.05M KOH	52
g-C$_3$N$_4$/Pt/ZnO	Hydrothermal and spin coating	120 μAcm^{-2} at 0.5 V$_{Ag/AgCl}$	0.5 M Na$_2$SO$_4$	54
Au/TiO$_2$(NTPC)	Photocatalytic reduction	150 μAcm^{-2} at 1.23 V$_{RHE}$	1 M KOH	59
Au/B-TiO$_2$	Hydrothermal and photodeposition	2.5 mAcm^{-2} at 1.23 V$_{RHE}$	0.5 M Na$_2$SO$_4$	18
Au/TiO$_2$ NT	Electrochemical anodization	1.02 mAcm^{-2} at 1.23V$_{RHE}$	1 M NaOH	62
1wt%-Au@TiO$_2$/Al$_2$O$_3$/Cu$_2$O	Hydrothermal and spin coating	-4.34 mAcm^{-2} at -0.2 V$_{Ag/AgCl}$	0.1 M Na$_2$SO$_4$	63
Au@SiO$_2$/ TiO$_2$	Drop casting	0.92 mA cm^{-2} at 0 V	0.5 M Na$_2$SO$_4$	64
Au/Fe$_2$O$_3$	*In situ* precipitation	2.0 mAcm^{-2} at 1.6 V$_{RHE}$	1 M KOH	65
Ag$_2$S/ TiO$_2$ TNA	Sequential-chemical bath deposition (S-CBD)	840 μA cm^{-2} at 0 V$_{Ag/AgCl}$	0.5 M Na$_2$S	66
Ag/ZnO	Electron beam evaporation	0.616 mAcm^{-2} at 1.23V$_{RHE}$	0.5 M Na$_2$SO$_4$	67
Pd/ TiO$_2$	Ultrasonic-assisted hydrothermal	3.6 μAcm^{-2} at -0.3 V$_{Ag/AgCl}$	0.1 M KCl	68
Pd/TiO$_2$ Oxynitride	Anodization	1.9 mAcm^{-2} at -0.894 V$_{Ag/AgCl}$	1.0 M KOH	69
Cu/TiO$_2$ TNAs	Pulsed electrochemical deposition	193 μA cm^{-2} at 0.2 V$_{Ag/AgCl}$	0.1 M NaNO$_3$	72

Table 2. Photocurrent measurements of noble bimetallic based photoanode in PEC water splitting

Photoelctrode	Metal loading method	Photocurrent density	Electrolyte	Ref.
Au-Pt/TiO$_2$	Drop-casting	$8\mu A\ cm^{-2}$ at0.75 $V_{Ag/AgCl}$	0.5 M Na$_2$SO$_4$	74
Au-Pd/TiO$_2$ NR	Cyclic voltammetry	3.36 mAcm^{-2} at 0.28 $V_{Ag/AgCl}$	0.1 M Na$_2$SO$_4$	20
Au-Pd/ZnO	Mild hydrothermal	0.98 mAcm^{-2} at 0.787 $V_{Ag/AgCl}$	0.5 M Na$_2$SO$_4$	76
ZnO@GR/Au-Pd	Electrostatic self-assembly and solution reduction	2.27mAcm^{-2} at 0.8 $V_{Ag/AgCl}$	0.5 M Na$_2$SO$_4$	81
Au-Ag/TiO$_2$	One-step co-sputtering	5.9 mAcm^{-2}at 0.5 $V_{Ag/AgCl}$	1 M KOH	57
Ag-Au/ TiO$_2$	Photodeposition	95 $\mu A\ cm^{-2}$ at 0 $V_{Ag/AgCl}$	1 M NaOH	82
Cu-Ni/ TiO$_2$	Sol-gel associated hydrothermal	2.29 mAcm^{-2} at 0.24 $V_{Ag/AgCl}$	0.5 M Na$_2$SO$_4$	85
Co-Fe/ TiO$_2$	Chemical bath deposition (CBD)	80 $\mu L\ cm^{-2}$ at 0.6 $V_{Ag/AgCl}$	1M KOH with 5vol% Ethylene gly--col solution	87
BiVO$_4$/TiO$_2$/FTO	Hydrothermal and chemical bath deposition	2.14 mAcm^{-2}at 1.23 V_{RHE}	0.5 M Na$_2$SO$_4$	88
ZNC-CH/TiO2	Hydrothermal and water- bath reaction	1.16 mAcm^{-2}at 1.23 V_{RHE}	0.25MNa$_2$SO$_4$	89
BiVO4/rGO/NiFe	Spin coating and electro-deposition	1.30 mAcm^{-2}at 1.23 V_{RHE}	0.5 M Na$_2$SO$_4$	90
CoO$_x$/SiO$_x$/n-Si	Atomic layer deposition (ALD)	3.5 mAcm^{-2}at 1.23 V_{RHE}	1 M KOH.	91
Au/CdS/LaBTC	Squeegee method	80 mAcm^{-2} at 0.8 $V_{Ag/AgCl}$	1 M Na$_2$S (P^H=13)	92
Au-Pd/ZnO NRs	Electrodeposition sputtering	0.345 mAcm^{-2} at 0.4 $V_{Ag/AgCl}$	0.1 M Na$_2$SO$_4$	77

4.1 Platinum (Pt)

Platinum, a well-known promising metal for photocatalytic water splitting reaction has been extensively used as photoanode in a photoelectrochemical cell [46]. Pt produces highest Schottky barrier among metals as it possesses highest work function value (5.6 eV) among the noble metals. This formation of Schottky junction with semiconductor can efficiently absorb the excited electrons from the conduction band (CB) of semiconductor suppressing the back flow of electrons. This phenomenon greatly improved the photocatalytic activity [47,48]. Fujishima and Honda developed the composite Pt/TiO_2 [4] in the year 1972 for an excellent photoelectrochemical water decomposition by direct light radiation without applying any electric power. In the experiment, the n-type TiO_2 was used as the anode and Pt was used as the cathode. When the TiO_2 electrode was irradiated in short circuit environment, H_2 gas was emitted at the Pt cathode and O_2 was emitted at the TiO_2 anode. Later on, the concept of this system was utilised by Bard to introduce a photocatalytic process [49] according to which the $Pt-TiO_2$ system recognised as a "short-circuited" photoelectrochemical cell, where a TiO_2 semiconductor electrode and a platinum-metal counter electrode have been connected. The role of metal particles was to act as mini-photocathode to absorb electrons therby reducing water to H_2 [50]. Timur Sh. Atabaev et al. successfully incorporated Pt nanoparticles on TiO_2 nanorods using a facile hydrothermal method [51] for performing photocatalytic performance in photoelectrochemical (PEC) water splitting. The photocurrent studies revealed that by tailoring the surface of TiO_2 with Pt nanoparticles, the photoelectrochemical activity of the TiO_2 photoelectrode can be greatly improved. To be precise, the Pt-doped TiO_2 nanorods have shown much higher photocurrent evolution than that of undoped ones. The reason for improved PEC activity of the Pt-doped TiO_2 NRs as compared to the undoped TiO_2 is the SPR effects of Pt nanoparticles and better charge separation mechanism on the interface between TiO_2 and Pt. The bare TiO_2 NRs have shown photocurrent density of 0.36 mA cm^{-2} (at 1.23 V vs. RHE) which upon deposition with Pt increased steeply by 86.1% to 0.67 mA cm^{-2}, indicating the promising role of Pt coating in enhancing the visible light photoactivity of TiO_2 [51].

Fangfang Wang and co-workers [52] demonstrated the effect of Pt loading on TiO_2 nanorods array thin film (TNTF) electrode in PEC water splitting to generate hydrogen. The hydrothermal method succeeding the sputtering method was employed to synthesize Pt/TiO_2 (TNTF), which exhibited an excellent photocurrent density with 3.8 mA/Cm^2 at 1.0 V$_{Ag/AgCl}$ which is nearly four- fold greater than that of annealed TNTF electrode with 0.96 mA/Cm^2. This improvement in photocatalytic activity was because of highly dispersed Pt NPs, which not only facilitated the photogenerated charge separation, but also increased the optical absorption in visible region effectively leading to increase photogenerated electrons [52].

Anna and Jerzy [53] reported the use of various sacrificial agents to improve hydrogen evolution capability of Pt/TiO$_2$ under ultraviolet (UV) irradiation and found much higher H$_2$ evolution rate of 11.50 mmol/90 min with methanol. The Pt played a very significant role in photocatalytic activity in the sense of drastic shifting of light absorption from UV to visible region [53].

In addition, synthesis of g-C$_3$N$_4$/Pt/ZnO as semiconductor-metal-semiconductor model by using hydrothermal method followed by spin coating has been reported by Jingran Xiao et al. [54]. They demonstrated that under AM 1.5 light illumination, the prepared C$_3$N$_4$/Pt/ZnO composite photoanode gave better PEC activity with greater photocurrent density of 120 μA cm^{-2} at 0.5 V vs. Ag/AgCl than that of the g-C$_3$N$_4$/ZnO. This difference in PEC performance was due to the presence of Pt which enhanced the visible light absorption, simultaneously acting as a co-catalyst to facilitate the transfer of photogenerated electrons through the suitable formation of Schottky junction. Also, g-C$_3$N$_4$ and ZnO played the role of light absorber to form electron-hole couple leading to improved PEC performance. In addition, Pt facilitated the photo induced electrons migration in g-C$_3$N$_4$ from the bulk to the exterior as well as accelerated the electron transfer from g-C$_3$N$_4$ to ZnO [54].

4.2   Gold (Au)

Although Pt is a beneficial metal, which enhances photoelectrode performance, it is costlier in comparison to gold. Gold possesses some unique properties including strong resistant to corrosion and surface plasmonic resonance (SPR) effect to induce hot electrons. Also, with the work function value of 5.1 eV, Au facilitates an appropriate formation of Schottky barrier between metal-semiconductor junction, which involves the trapping of electrons from semiconductor, as a consequence charge separation becomes prominent which in turn help in improvising photocatalytic H$_2$ evolution [55]. There are several reports on Plasmonic induced photoelectrochemical water splitting. Ueno and Misawa [56] studied Plasmon induced photocurrent development and oxidation of H$_2$O on Au nanorods (NRs) assembled on TiO$_2$ surface. H$_2$ and O$_2$ evolutions were obtained with reduced over potential in the photoelectrochemical cell. No current was observed when bare TiO$_2$ was illuminated under visible or near-IR radiation, whereas anodic photocurrent was obtained at a positive potential above -0.3V$_{Ag/AgCl}$ with the Au NRs deposited TiO$_2$ photoelectrode. The excited electrons were incorporated in the conduction band (CB) of TiO$_2$ due to photoexcitation of plasmonic Au NPs and the holes left behind were involved in the oxidation of water, therefore, photocurrent density was increased in the case of Au incorporated TiO_2 nanostructures [20,56]. Similar observation was noticed by Haro et al. [57] on adding Au NPs to mesoporous TiO_2 on the conducting substrate in which the plasmonic Au NPs played a crucial role to improve photocatalytic performance [57]. Kim et al. [58] found that the photocurrent generation was remarkably influenced by plasmonic property and controllable size of Au NPs. The plasmonic impact in the photoelectrochemical water splitting reaction was more prominent with smaller Au particle size. They found much larger photocurrent for 50 nm Au NPs and on increasing the Au NPs size photocurrent was reduced. The reason for the improved photoelectrochemical water splitting activity on decreasing gold particle size was attributed to the increase in charge carrier generation on decreasing the size [58].

The plasmonic incorporated photonic crystals have been widely used in photoelectrochemical water splitting. Photonic crystals (PCs) are known to trap light within their periodic structures, thereby, slowing down the photons [59,60]. Zhang et al. [59] reported the deposition of plasmonic Au nanocrystal on TiO_2, which exhibited superb photoelectrochemical water splitting activity with photoconversion rate of 1.1%. The photocurrent generation of 150 μA cm^{-2} at 1.23 V_{RHE} potential was observed for Au/TiO_2 photonic crystal (PC) composite which is around 50 fold larger than that of TiO_2 nanotube photonic crystal (PC). However, without a PC layer the evolved photocurrent densities for TiO_2 and Au/TiO_2 nanotubes were found to be 3 and 60 μA cm^{-2} respectively. This improved PEC performance in presence of photonic crystal (PC) was because of the matching of surface plasmonic resonance (SPR) of Au NPs with photonic band gap of the PC and hence photonic crystal substrate has significant impact on the SPR of Au.

The uniform deposition of Au NPs on branched TiO_2 array using hydrothermal followed by photo-deposition method has been well demonstrated by Fang Xu et al. [18] who observed an excellent activity with photoconversion efficiency of 0.6%. The Au/B-TiO_2 composite showed an improved PEC performance with photocurrent density of 2.5 mA cm^{-2} at 1.23 V_{RHE} potential, which is comparatively very high than those of B-TiO_2 and TiO_2 nanorods. This outstanding PEC activity is owing to improved solar radiation capturing ability, improved charge carrier density which is an outcome of the synergistic impact of highly-branched structure of TiO_2 and surface plasmon resonance (SPR) effect of Au NPs. Often, this photoanode has been beneficial in various potential utilizations in energy storage and conversion fields [18,61].

Besides, inculcation of Au NPs on TiO_2 nanotube (TNT) template with a simple technique involving of double-step anodization and Au thin film accumulation has proven to be useful because of tremendous PEC activity with photocurrent density of 1.02 mA cm^{-2} at 1.23 V_{RHE}. This enhanced PEC performance of the TNT array can be attributed to charge separation and transportation for reduced cracking following the second anodization and the annealing method [62]. Piangjai and co-workers [63] have reported the fabrication of p-n heterojunction photoelectrode using Cu_2O and TiO_2-P25 deposited with only 1 wt% Au@TiO_2 plasmonic core-shell structure by hydrothermal and spin-coating processes to be used in photoelectrochemical (PEC) water splitting. The introduction of that plasmonic core-shell structure remarkably improved light harvesting and stimulated charge carrier separation, which significantly enhanced photocurrent density up to -4.34 mA cm^{-2} at -0.2 $V_{Ag/AgCl}$ by using simulated illumination, that is around 20 times higher in comparison to that of TiO_2- P25/Cu_2O photoelectrode. The core shell design of Au NPs, size of Au NPs and SPR induced plasmonic absorption of Au NPs have considerable impact in improving PEC performance. Moreover, the multilayer photoelectrodes have shown elevated stability by systematic fabrication of the PEC system to circumvent the risk of photocorrosion on Cu_2O layers [63].

In addition, Jianan Chen et al. [64] have reported another core-shell of Au-SiO_2 NPs deposited in the exterior of TiO_2 NRs prepared through a simplistic double-step drop casting process. The synthesized Au-SiO_2/TiO_2 NR films showed remarkably enhanced photocurrents relative to the pristine TiO_2 NR film. The Au@SiO_2 core shell NP-decorated TiO_2 nanorod film has remarkably improved photocurrent density from 0.60 to 0.92 mA/cm^2 and incident photon to current efficiency (IPCE) at 380 nm was increased from 31% to 37%. The photoelectrochemical activity enhancement was deductively induced by the high-intensity electromagnetic field originated from the metallic plasmonic resonance, which effectively promoted the charge carriers separation formed in the proximal surface region of TiO_2 nanorods [64].

Lei Wang et al. [65] have successfully demonstrated much better photoelectrochemical (PEC) water splitting photocataytic activity of Au NPs decorated haematite nanoflake template relative to the undoped Fe_2O_3. The Au-Fe_2O_3 photoelectrode showed a notable cathodic shift of the initial potential up to 0.6 V_{RHE} and a two-fold increase in the water oxidation photocurrent at 1.23 V_{RHE}. This improved photocurrent was due to the Au NPs serving as plasmonic photo sensitizers which augmented the optical absorption [65].

4.3   Silver (Ag)

Silver is one of the precious metals having the work function of 4.26 eV and facilitating the formation of Schottky barrier. These two properties of Ag such as Schottky barrier and SPR induced plasmonic absorption of Ag NPs have been widely utilized to improve photoelectrochemical hydrogen evolution. The plasmonic enhanced photoelectrochemical water splitting by Ag_2S NPs deposited on TiO_2 using sequential-chemical bath deposition (S-CBD) process has been widely studied by Mostafa Gholami and co-workers [66].

Two-step anodization process used to prepare TiO_2 nanotube array (TNA) has supported the formation of TNA/Ag_2S-n nanostructure in which TNAs were photoelectrochemically activated by Ag_2S NPs. The most excellent photocurrent measurement was found with 840 $\mu A/cm^2$ which is around 15 fold more than that obtained for bare TNA. Here Ag_2S NPs covered the charge recombination centers and the improvised electron–hole couple separation boosted the PEC activity. In fact, the incident photon to current efficiency (IPCE) measurements revealed that the visible response of the materials obtained the peak value of around 20% with a wavelength of 600 nm for the TNA/Ag_2S nanostructure [66].

Another work reported by Yuefan Wei et al. [67] demonstrated the ZnO NR fabrication using Ag coat on the polyethylene terephthalate (PET) template through a low-temperature hydrothermal process and electron beam evaporation acting as an outstanding photoanode for water splitting. The plasmonic stimulated absorption, surface recombination hindrance and enhanced charge transport impact were studied in different Ag thickness and found that the prepared substrate with 10 nm thickness bent to R = 6.0 mm attained much better light trapping than that of the substrate without bending. The maximum photocurrent with 0.616 mA/cm^2 and the photoelectron conversion efficiency (PCE, 0.81%) were obtained. The efficient PEC activity was due to the plasmonic effects of the Ag film and better charge transport due to charge recombination hindrance of ZnO surface. Excellent light harvesting stimulated by PET substrates bending promoted the overall water splitting efficiency [67].

4.4   Palladium (Pd)

The incorporation of Pd to modify the semiconductor materials has received much attention in photocatalytic field to improve PEC performance. Pd improves the plasmonic absorption as well as electron-hole pair recombination period of nanocomposite. Junjie Wu et al. [68] developed a novel Pd/TiO_2 nanosheets photocatalyst by depositing Pd NPs onto the TiO_2 nanosheets with a large extent of exposed (001) facets through a simple chemical reduction method. The Pd/TiO_2 nanosheets with varying amounts of Pd loadings, in relative to bare TiO_2 showed better catalytic efficiency in hydrogen evolution from photocatalytic water splitting under light illumination of 300 W Xenon lamps. The Pd/TiO_2 nanosheet catalyst containing 0.18 wt% Pd NPs has shown the highest photocatalytic performance with hydrogen evolution of 3096 $\mu molg^{-1}h^{-1}$ and photocurrent generation of 3.6 $\mu A\ cm^{-2}$. The suitable work function to form appropriate Schottky junction and plasmonic absorption of Pd as well as synergistic interaction between TiO_2 nanosheets and Pd have played vital roles in improving the photocatalytic water splitting efficiency [68].

Nageh K. Allam and co-workers [69] have demonstrated a photoanode component prepared by self-structured, vertically directed nanotube substrate of titanium-palladium incorporating oxynitride films by anodization of Ti-Pd alloy in an formamide electrolyte solution with NH_4F in it at room temperature and then by annealing in an ammonical environment. The Ti-Pd oxynitride nanotube was used in photoelectrochemical water splitting, giving a photocurrent density of 1.9 mA/cm^2, which is far better than that of bare TiO_2 nanotube. This improvement in the photoconversion efficiency can be attributed to the synergistic effects of Pd and nitrogen doping and also the exceptional structural properties of the synthesized nanotubes. In fact, oxynitrides served a major function in uprising the catalytic activity because of more narrow band gaps in comparison to analogous metal oxides. The metal- nitrogen bond in oxynitrides has a more potential energy relative to the metal-oxygen bond in metal oxides. Besides, oxynitrides are stable enough against anodic dissolution in basic environment, and hence are better for water splitting [69].

4.5   Copper (Cu)

Among the first row of transition metals, Cu is sometime kept with the group of noble metals based on the completely filled electronic configuration of d-orbital. Copper exhibits some unique properties such as surface plasmonic resonance (SPR), electrical conductivity and photocatalytic properties for which Cu has immensely been utilized in the field of photoelectrochemical (PEC) water splitting. Moreover, use of Cu reduces the cost compared to the expensive noble metals. Also, the work function of Cu with 4.65 eV widely influences the PEC performance of the composites [42,70,71]. Zhang et al. demonstrated [72] the Cu NPs loaded TiO_2 arrays (TNAs) photoanode (Cu/TNAs) for photoelectrochemical water splitting and also photocatalytic H_2 evolution in visible radiation. The synthesized Cu/TNAs executed the photocurrent generation of 193 $\mu A/cm^2$ at 0.2 V vs. Ag/AgCl, which is 11 times higher in comparison to bare TNAs. Besides, the photocatalytic hydrogen evolution rate attained from Cu/TNAs was larger enough relative to bare TNAs, because of the excellent SPR stimulated plasmonic absorption of the Cu NPs in Cu/TNAs. The inherent plasmonic SPR properties of Cu increase the visible light absorption of photoanode generating large number of electron-hole pairs which consequently increase the photocatalytic performance in photoelectrochemical water splitting. In addition, Cu exhibited an efficient electron trapping through the formation of an appropriate Schottky barrier with semiconductor leading to suppress the charges recombination which in turn improves hydrogen emission from photoelectrochemical (PEC) water splitting. The hot electrons stimulated by plasmon are introduced partially in the conduction band (CB) of TiO_2 above the Schottky junction since the work function of Cu (4.6 eV) is relatively more negative as compared to TiO_2

(4.9 eV) because of which the electron-hole pairs in Cu NPs are divided easily. The high energy electrons react with H^+ at the TNA surface to produce H atom, and then they assemble amongst them to form H_2 molecules. This phenomenon is supporting the enhancement in photocatalytic hydrogen production [72,73].

5.   Noble bimetallic nanocomposites for PEC water splitting

5.1   Au-Pt bimetallic nanocomposites

There are several noble bimetallic alloys which have been efficiently used in photoelectrochemical (PEC) water splitting. Since, platinum (Pt) and gold (Au) NPs deposited on photoelectrode delivered excellent photocatalytic performance under light irradiation, combination of these two metals remarkably improved their photocatalytic performance for H_2 evolution reactions. This enhanced photocatalytic performance of bimetallic alloy can be explained from the two point of views. One is improved visible radiation absorption of the photoelectrode from the synergistic LSPR effect shown by the deposited metals and another view is due to formation of bimetallic alloy the work function gradually decreases forming an appropriate Schottky barrier between metals-semiconductor interfaces enhancing the suppression of electron-hole pair recombination and hence improving the PEC activities. Fenglong Wang et al. [74] have developed the size reliant Au NPs synthesized by a seeded growth process for the purpose of sensitization of Pt/TiO_2 as visible-light receptive catalysts for hydrogen evolution. Photoelectrochemical (PEC) measurements showed that among all the analysed catalysts, Au_{20}-Pt/TiO_2 displayed exceptionally well hot electrons separation, associated best with the photocatalytic activity. The finest performance of Au_{20}-Pt/TiO_2 evidently shows the dependence on size of the surface plasmon resonance of Au NPs and enhanced activity relative to Au_{20}/TiO_2 signifies the imperative function of Pt as co-catalyst for proton reduction. Pt with its well-arranged Schottky junction can effortlessly entrap the photogenerated hot electron on Au NPs by the conduction band (CB) of TiO_2 which leads to photocatalytic reaction. Moreover, Pt^{6+} can be reduced simultaneously on TiO_2 as well as Au NPs surfaces by using Au_{20}/TiO_2 in visible radiation, supporting the fact that the proton reduction reaction can take place together on the excited Au NPs surface and TiO_2 supports. Afterward, studies were performed on charge transfer mechanism through PEC measurements using various sacrificial agents like methanol and ethanol as holes scavengers. While using methanol, the PEC performance of $Au_{20}Pt/TiO_2$ gave about 50% more photocurrent response relative to ethanol, a weaker hole scavenger. The holes scavenging dependence of the photocurrent response reveals that the SPR is improved for the electron-hole charge transfer mechanism [74].

In addition, Ambrose A. Melvin and coworkers [75] demonstrated the preparation of Au/TiO_2 as the substrate composite, on which Au along with other metals (Ag, Pt or Pd) was coated concurrently using a simplistic one pot photodeposition process under an optimum composition and the resulted system was investigated for solar water splitting (SWS) in the presence as well as absence of a sacrificial agent. The maximum H_2 yield was obtained for $Pt_{0.5}$-Au/TiO_2 is 1275 $\mu molg^{-1}h^{-1}$, succeeded by $Pd_{0.5}$-Au/TiO_2 872 $\mu molg^{-1}h^{-1}$ and $Ag_{0.5}$-Au/TiO_2 466 $\mu molg^{-1}h^{-1}$ using methanol as sacrificial agent. The same composite was also found highly active for SWS in absence of sacrificial agent. The improved photocatalytic performance was because of plasmonic absorption and proficient electron trap by Au and Pt by the development of Schottky junction with semiconductor. To uprise the H_2 evolution, the content of noble metal has to be increased but without growing the particle size. Because of this the charge carrier recombination minimized and charge carriers utilization boosted up. However, the distance between redox sites and charge generation sites diminished [75].

5.2   Au-Pd bimetallic nanocomposites

The surface modification of the photoelectrode with bimetallic Au-Pd alloy has resulted in remarkable improvement in the efficiency of photoelectrochemical (PEC) water splitting for hydrogen generation. The enhanced photocatalytic performance has been attributed to the presence of metal clusters or alloy on titania, which are supposed to promote the separation of photogenerated charges and hence more electrons and holes can be involved in reduction and oxidation process. Therefore, in H_2 evolution reactions the alloy induced the synergistic effect and improved the photocatalytic performance of nanocomposites.

The fabrication of TiO_2 nanorods with bimetallic Au-Pd NPs without using any surfactant or polymer has been reported by Roozbeh Siavash Moakhar et al. [20]. The resulted nanorods exhibited high photocatalytic performance for photoelectrochemical (PEC) water splitting yielding outstanding photoconversion efficiency of 2.31% in AM 1.5 irradiation compared to other designed TiO_2 1D nanostructures. The photoelectrode Au-Pd/TiO_2 showed a photocurrent density of 3.36 mA/cm^2. EIS and Mott–Schottky plots also revealed that Au-Pd/TiO_2 having lowest charge transfer resistance and highest carrier density has faster carrier transfer. Furthermore, Au-Pd NPs assist the overall enhancement of PEC activity by generating a Schottky junction, inhibiting electron–hole recombination and passivate electron trapping of TiO_2 NRs which ultimately boost the photocurrent drastically [20].

Also, the excellent photoelectrochemical (PEC) performance of 1D ZnO decorated Au-Pd bimetallic NPs through a mild hydrothermal method has been reported by Yan Lu and co-workers [76]. They investigated the optical properties of Au-Pd/ZnO by PEC studies and found the photocurrent generation up to 0.98 mA/cm^2 at 0.787 V versus Ag/AgCl, which indicated the better photoactivity compared to the bare ZnO NR substrate in photoelectrochemical H$_2$ evolution reaction. The improvement in catalytic performance was attributed to the higher photocatalytic activity of Au-Pd bimetallic NPs because of the plasmonic properties of Au and the finer catalytic behavior of Pd which energised the electron on the Au-Pd alloy [77]. The second reason may be that the Au-Pd alloy NPs deposited on the ZnO NR surface can increase the light utilization efficiency through reflection of unabsorbed photons to the ZnO NR substrate [78,79]. Another reason would be the role of ZnO NR developed on the conductive glass substrate that can directly carry the energized electrons to external electrode [80]. Thus, the photoactivity of Au-Pd/ZnO composite considerably increased because of the synergistic effect between 1D ZnO nanorod arrays and Au-Pd alloy NPs. The insertion of graphene layer to the composite has significant impact on photocatalytic H$_2$ evolutions from PEC water splitting.

Yuzhi Zhang et al. [81] fabricated Au-Pd NPs in graphene-layered ZnO nanorod (ZNR) electrodes (ZNR@Gr/Au-Pd) by electrostatic self-assembly and solution reduction process. The composite exhibited an excellent PEC performance and stability with photocurrent density of 2.27 mA/cm^2 at 0.8 V against Ag/AgCl, that is 7.1 fold more than that of pure ZNR, displaying noteworthy prospective in hydrogen generation. The reason behind this excellent PEC activity and stability is the LSPR induced excited electrons on Au-Pd immediately transfer to graphene as the Fermi level of graphene (-4.42 eV) is lesser than the conduction band (CB) of ZnO. Consequently, the photoexcited interior and exterior electrons of ZnO are accumulated in graphene which are immediately transported to the Pt electrode. Furthermore, the exterior graphene layering on the ZnO NR could hinder the developed Zn^{2+} ions from diffusion into the electrolyte therby protecting it from photocorrosion [81].

In our work, we have studied on photocatalytic H$_2$ evolutions by noble bimetallic alloy deposited on photoelectrode TiO$_2$. A mild hydrothermal process was employed to prepare the Au-Pd/rGO/TiO$_2$ composite, which exhibited an excellent photocatalytic activity with H$_2$ evolution rate of 21.49 mmolh^{-1} with methanol as a sacrificial agent. The photocatalytic performances of the composite in the forms of thin film and powder were evaluated and it was found that the thin film form produced much better performance than corresponding powder form which can be explained as the bimetallic NPs enhanced the visible light absorption through LSPR effect producing photoexcited electrons, which immediately transfer to the conduction band (CB) of TiO$_2$ through reduced graphene oxide (rGO). Then, it is the Pd co-catalyst which immediately entraps the electrons in the CB of TiO$_2$ reducing the photogenerated electron-hole recombination through formation of an appropriate Schottky barrier between metal-semiconductor interfaces. This composite showed much higher photocatalytic activity compared to bimetallic Au-Pd NPs deposited on TiO_2 with H_2 evolution rate of 6.16 mmol $g^{-1}h^{-1}$, clearly indicating the electrons mediator role of reduced graphene oxide (rGO) assisting the inhibition of recombination of the electron-hole pairs and consequently enhanced the photocatalytic performance.

5.3   Au-Ag bimetallic nanocomposites

The bimetallic Au-Ag nanocomposites are promising materials for H_2 evolutions from water splitting. Presence of these noble Au-Ag bimetallic NPs on photoelectrode greatly improves the photoelectrochemical (PEC) performance through utilization of their dual plasmonic properties. Considering hydrogen as an attractive and reliable future fuel, Naimeh Naseri and co-workers [57] have reported an efficient method for production of H_2 by utilizing Au-Ag/TiO_2 composite in photoelectrochemical (PEC) water splitting. A one-step co-sputtering method was utilized to stack Au-Ag NPs on TiO_2, which showed excellent PEC activity with photocurrent density increased from 4.6 to 5.9 mA/cm^2 indicating the influence of plasmonic property of Au-Ag NPs. Photocurrent studies illustrated a 30% rise with the alloy NPs, and a 50% reduction in charge transfer hindrance of the electrodes. Some reports to demonstrate the annealing ambient effect have also appeared. TiO_2 and Au-Ag/TiO_2 NPs were allowed to anneal in air (ambient oxygen) under similar temperature condition and the photoresponses of both entities were determined under similar electrochemical environment. The photocurrent density of the photoanodes was found as 0.7 for oxygen-treated TiO_2 and 1.3 mA/cm^2 for Au-Ag/TiO_2 films. Thus, it can be appropriately said that the reducing atmosphere, which paved the way for oxygen vacancies in the systems has significant impact in rising the photoresponses, while both the samples being in annealing conditions, inclusion of Au–Ag alloy NPs also hike the PEC activity of the systems [57].

Another work by Kshirodra K. Patra et al. [82] demonstrated a development of a prospective method for utmost consumption of the complete visible region of the solar illumination by devising a plasmonic metals semiconductor composite. Ag together with Au nanoclusters would extend the visible light absorption at least between 400 and 650 nm. Solar water splitting (SWS) performances of Au–TiO_2, Ag–TiO_2 and Ag on Au–TiO_2 composites have been studied and found that Ag-Au/TiO_2 bimetallic nanocomposite delivered maximum perceptible quantum yield of 3.3% and hydrogen evolution rate of 718 mmolg$^{-1}h^{-1}$ using aqueous methanol. The Photocurrent measurements also showed considerable enhance in the current density after depositing Ag onto Au/TiO_2. Compared to TiO_2 (25 µA/cm^2) and Au/TiO_2 (58 µA/cm^2), 40Ag-Au/TiO_2 photoanode attained the maximum photocurrent deliveration of 95 $\mu A/cm^2$ at zero bias. The drastic increase (almost twice compared to Au/TiO_2) of photocurrent generation by mere doping Ag on Au/TiO_2 reflects the efficient light harvesting capacity by the blend of Ag and Au. This displays the optimistic function of Ag NPs in improving the photoresponse of 40Ag-Au/TiO_2 composition signifying a superior light absorption ability, the reason of which may be due to generation of hot electrons. Thus, it can be concluded that Fermi level equilibrium existing between metal and TiO_2 and Schottky barrier formation are the prime reasons for efficient photocatalytic performance [82].

5.4   Ag-Cu bimetallic nanocomposites

Presence of relatively less expensive noble bimetallic Ag-Cu NPs at the photoelectrode surface significantly improve the efficiency of photoelectrochemical (PEC) water splitting. Ag and Cu both exhibited SPR induced plasmonic absorption producing more number of hot electrons. Liu and co-workers [83] developed a plasmonic composite of Ag-Cu NPs onto TiO_2 nanotube arrays (TNAs) by using photodeposition method. The developed Ag-Cu/TiO_2 composite with molar proportion of 1: 4 (Ag: Cu) gave elevated short-circuit photocurrent than Ag/TNAs and Cu/TNAs in both simulated sunlight as well as visible irradiation. The synergistic plasmonic absorption resulted from Ag-Cu alloy played a crucial role in improving the PEC activity. The achieved photocurrent generation of Ag-Cu/TNAs (1:4) in simulated sunlight and visible light were -1.201 and -0.734 mA/cm^2 respectively, assisted by the enhanced photogenerated electrons [83].

6.   A brief note on bimetallic non-noble NPs for photoelectrochemical (PEC) water splitting

Although noble metals are very effective for PEC water splitting, but are more expensive. Non-noble metals can be an alternative towards enhancing PEC activities. The transition-metals (Ni, Co and Fe) supported materials as well as the metals alone and their alloys have extensive applications in PEC water splitting for H_2 and O_2 evolutions because of their cost-effectiveness, high abundance and exceptional catalytic features. The plasmonic effect based application in PC and PEC water splitting of these three metals has very few reports. This might be due to their moderately fragile plasmonic resonance property resultant from the substantial damping [84]. Thus, metallic Ni, Co and Fe supported materials have been broadly used as co-catalysts to effectively split water. In comparison to metallic Co and Fe, alloys based on Ni have been extensively applied by depositing on semiconductor surface for water reduction half reaction. A study from Norani Muti Mohamed et al. [85] reported that the loading of Cu-Ni nanoalloy on TiO_2 using   sol-gel   associated   hydrothermal   method   exhibited   an   excellent

Photoelectrochemical Water Splitting: Materials and Applications
Materials Research Foundations **70** (2020) 183-213

Materials Research Forum LLC
https://doi.org/10.21741/9781644900734-7

photoelectrochemical (PEC) water splitting behavior compared to monometallic Ni and Cu deposited TiO_2 and pure TiO_2 photocatalysts. The photoelectrochemical measurement revealed that Cu-Ni (5:5) deposited TiO_2 thin film generated maximum hydrogen about 5.3 mL from solar water splitting with photocurrent formation of 2.29 mA/cm^2 at 0.24 V and outstanding photoconversion efficiency of 4.33%. The reason for the superior PEC performance relative to other photocatalysts is the synergetic effect of both metals as charge carriers trapper, large electron-hole pair separation, high negative flat band potential to split water and elevated electron donor density. Since, metallic copper is SPR active inducing the plasmonic absorption leading to huge generation of electrons on conduction band (CB) of semiconductor and consequently transfer of these electrons occurs to Schottky barrier, which is formed by Cu-Ni alloy due to an appropriate work function. This mechanism significantly improves the photocatalytic performance to generate hydrogen [57,86]. Mohamad Mohsen Momeni with co-workers [87] synthesized TiO_2 nanotubes by using electrochemical oxidation method followed by loading of Co and Fe NPs through chemical bath deposition method (CBD). The synthesized nanocomposites were used as photoanodes in photoelectrochemical (PEC) water splitting for hydrogen generation. The achieved hydrogen evolution was 80 μL cm^{-2} at bias potential of 0.6 V against Ag/AgCl in a basic electrolyte of 1 M KOH consisting 5 vol% ethylene glycol solutions. Here, the metallic Co and Fe have great impact on photocatalytic water splitting reaction. The beneficial properties of these photoanodes include cost-effectiveness, high performance under light radiation and high stability [87].

Conclusion

Various promising semiconductors based noble metals photoelectrodes have been discussed for effective solar to hydrogen (STH) energy conversion. Wide band gap as well as rapid electron-hole pair recombination have been the challenging causes responsible for low solar energy conversion efficiency of semiconductor in PEC reactions. The electronic properties of the photoelectrodes can be smoothly tuned through the incorporation of noble metals. The noble metals efficiently improve the solar light absorption owing to association of surface plasmon resonance (SPR) property which in turn helps in generation of increased photogenerated hot electrons. These electrons are easily trapped and transport through Schottky barrier formed from metals-semiconductors interface to the reaction surface initiating the process of hydrogen evolution reactions (HER) whereas the holes generated inside involve in oxygen evolution reactions (OER). Although, the overall photoelectrochemical (PEC) cell for water splitting has still low efficiency, it is now indispensable to work for the enhancement of the efficiency and realistic applicability of the photocatalytic water splitting system predominantly the advancement of highly efficient and low-cost materials. The distinctive properties of noble materials have significant key role in developing systems having high photocatalytic efficiency for photoelectrochemical water splitting utilising solar radiation.

References

[1] T. Jafari, E. Moharreri, A.S. Amin, R. Miao, W. Song, S.L. Suib, Photocatalytic water splitting-the untamed dream: a review of recent advances, Molecules 21 (2016) 900-929. https://doi.org/10.3390/molecules21070900

[2] P. Cheng, Z. Yang, H. Wang, W. Cheng, M. Chen, W. Shangguan, G. Ding, TiO_2-graphene nanocomposites for photocatalytic hydrogen production from splitting of water, Int. J. Hydrog. Energy 37 (2012) 2224-2230. https://doi.org/10.1016/j.ijhydene.2011.11.004

[3] P. Peerakiatkhajohn, J.-H. Yun, S. Wang, L. Wang, Review of recent progress in unassisted photoelectrochemical water splitting: from material modification to configuration design, J. Photon. Energy 7 (2016) 012006-012021. https://doi.org/10.1117/1.JPE.7.012006

[4] A. Fujishima, K. Honda, Electrochemical photolysis of water at a semiconductor electrode, Nature 238 (1972) 37-38. https://doi.org/10.1038/238037a0

[5] J. Li, N. Wu, Semiconductor based photocatalysts and photoelectrochemical cells for solar fuel generation: a review, Catal. Sci. Technol. 5 (2015) 1360-1384. https://doi.org/10.1039/C4CY00974F

[6] A.J. Bard, Photochemistry and heterogeneous photocatalysis at semiconductors, J. photochem. 10 (1979) 59-75. https://doi.org/10.1016/0047-2670(79)80037-4

[7] J. Cen, Q. Wu, M. Liu, A. Orlov, Developing new understanding of photoelectrochemical water splitting via *in-situ* techniques: a review on recent progress, Green Energy & Environ. 2 (2017) 100-111. https://doi.org/10.1016/j.gee.2017.03.001

[8] J. Joy, J. Mathew, S.C. George, Nanomaterials for photoelectrochemical water splitting -review, Int. J. Hydrog. Energy 43 (2018) 4804-4817. https://doi.org/10.1016/j.ijhydene.2018.01.099

[9] Y. Tachibana, L. Vayssieres, J.R. Durrant, Artificial photosynthesis for solar water splitting, Nat. Photonics 6 (2012) 511-518. https://doi.org/10.1038/nphoton.2012.175

[10] J. Brillet, J.H. Yum, M. Cornuz, T. Hisatomi, R. Solarska, J. Augustynski, M. Graetzel, K. Sivula, Highly efficient water splitting by a dual absorber tandem cell, Nat. Photonics 6 (2012) 824-828. https://doi.org/10.1038/nphoton.2012.265

[11] S. Ikeda, T. Itani, K. Nango, M. Matsumura, Overall water splitting on tungsten-based photocatalysts with defect pyrochlore structure, Catal. Lett. 98 (2004) 229-233. https://doi.org/10.1007/s10562-004-8685-y

[12] L. Meda, L. Abbondanza, Materials for photo-electrochemical water splitting, Rev. Adv. Sci. Eng. 2 (2013) 200–207. https://doi.org/10.1166/rase.2013.1034

[13] A. Wolcott, W.A. Smith, T.R. Kuykendall, Y. Zhao, J.Z. Zhang, Photoelectrochemical water splitting using dense and aligned TiO_2 nanorod arrays, Small 5 (2009) 104-111. https://doi.org/10.1002/smll.200800902

[14] V. Cristino, S. Caramori, R. Argazzi, L. Meda, G.L. Marra, C.A. Bignozzi, Efficient photoelectrochemical water splitting by anodically grown WO_3 electrodes, Langmuir 27 (2011) 7276-7284. https://doi.org/10.1021/la200595x

[15] Y. Ling, G. Wang, D.A. Wheeler, J.Z. Zhang, Y. Li, Sn-doped hematite nanostructures for photoelectrochemical water splitting, Nano Lett. 11 (2011) 2119-2125. https://doi.org/10.1021/nl200708y

[16] F. Su, J. Lu, Y. Tian, X. Ma, J. Gong, Branched TiO_2 nanoarrays sensitized with CdS quantum dots for highly efficient photoelectrochemical water splitting, Phys. Chem. Chem. Phys. 15 (2013) 12026-12032. https://doi.org/10.1039/c3cp51291f

[17] C.-Y. Lin, Y.-H. Lai, D. Mersch, E. Reisner, Cu_2O/NiO_x nanocomposite as an inexpensive photocathode in photoelectrochemical water splitting, Chem. Sci. 3 (2012) 3487-3489. https://doi.org/10.1039/c2sc20874a

[18] F. Xu, J. Mei, M. Zheng, D. Bai, Z. Wu, K. Gao, K. Jiang, Au nanoparticles modified branched TiO_2 nanorod array arranged with ultrathin nanorods for enhanced photoelectrochemical water splitting, J. Alloys Compd. 693 (2017) 1124–1132. https://doi.org/10.1016/j.jallcom.2016.09.273

[19] S.T. Kochuveedu, Photocatalytic and photoelectrochemical water splitting on TiO_2 via photosensitization, J. Nanometer. 2016 (2016) 1-12. https://doi.org/10.1155/2016/4073142

[20] R. siavash Moakhar, M. Jalali, A. Kushwaha, G. Kia Liang Goh, N. Riahi-Noori, A. Dolati, M. Ghorbani, Au-Pd bimetallic nanoparticle decorated TiO_2 rutile nanorod arrays for enhanced photoelectrochemical water splitting, J. Appl. Electrochem. 48 (2018) 995–1007. https://doi.org/10.1007/s10800-018-1231-1

[21] P. Aurora, P. Rhee, L. Thompson, Titania nanotube supported gold photoanodes for photoelectrochemical cells, J. Electrochem. Soc. 157 (2010) 152-155. https://doi.org/10.1149/1.3417096

[22] C. Langhammer, Z. Yuan, I. Zoric, B. Kasemo, Plasmonic properties of supported Pt and Pd nanostructures, Nano Lett. 6 (2006) 833–838. https://doi.org/10.1021/nl060219x

[23] A. Zieli´nska-Jurek, E. Kowalska, J.W. Sobczak, W. Lisowski, B. Ohtani, A. Zaleska, Preparation and characterization of monometallic (Au) and bimetallic (Ag/Au) modified-titania photocatalysts activated by visible light, Appl. Catal. B: Environ. 101 (2011) 504–514. https://doi.org/10.1016/j.apcatb.2010.10.022

[24] T. Pakizeh, C. Langhammer, I. Zoric, P. Apell, M. Kall, Intrinsic fano interference of localized plasmons in Pd nanoparticles, Nano Lett. 9 (2009) 882–886. https://doi.org/10.1021/nl803794h

[25] S.E. Hunyadi Murph, K.J. Heroux, C.E. Turick, D. Thomas, Metallic and hybrid nanostructures: fundamentals and applications, J. Nanosci. Nanotech. 4 (2013) 387-427.

[26] A. Zielińska-Jurek, Progress, challenge and perspective of bimetallic TiO$_2$ based photocatalysts, J. Nanomater. 2014 (2014) 1–17. https://doi.org/10.1155/2014/208920

[27] E. Kowalska, H. Remita, C. Colbeau-Justin, J. Hupka, J. Belloni, Modification of titanium dioxide with platinum ions and clusters: application in photocatalysis, J. Phys. Chem. C 112 (2008) 1124–1131. https://doi.org/10.1021/jp077466p

[28] A. Zieli´nska-Jurek, J. Hupka, Preparation and characterization of Pt/Pd-modified titanium dioxide nanoparticles for visible light irradiation, Catal. Today 230 (2014) 181–187. https://doi.org/10.1016/j.cattod.2013.09.045

[29] Y. Shiraishi, Y. Takeda, Y. Sugano, S. Ichikawa, S. Tanaka, T. Hirai, Highly efficient photocatalytic dehalogenation of organic halides on TiO$_2$ loaded with bimetallic Pd-Pt alloy nanoparticles, Chem. Commun. 47 (2011) 7863–7865. https://doi.org/10.1039/c1cc12087e

[30] Y. Mizukoshi, K. Sato, T.J. Konno, N. Masahashi, Dependence of photocatalytic activities upon the structures of Au-Pd bimetallic nanoparticles immobilized on TiO$_2$ surface, Appl. Catal. B: Environ. 94 (2010) 248–253. https://doi.org/10.1016/j.apcatb.2009.11.015

[31] L. Li, Z. Xu, F. Liu, Y. Shao, J. Wang, H. Wan, S. Zheng, Photocatalytic nitrate reduction over Pt–Cu/TiO$_2$ catalysts with benzene as hole scavenger, J. Photochem. Photobiol. A 212 (2010) 113–121. https://doi.org/10.1016/j.jphotochem.2010.04.003

[32] C.-T. Hsieh, J.-Y. Lin, Fabrication of bimetallic Pt-M (M= Fe, Co and Ni) nanoparticle/carbon nanotube electrocatalysts for direct methanol fuel cells, J. Power Sources 188 (2009) 347–352. https://doi.org/10.1016/j.jpowsour.2008.12.031

[33] Y. Ji, S. Yang, S. Guo, X. Song, B. Ding, Z. Yang, Bimetallic Ag/Au nanoparticles: a low temperature ripening strategy in aqueous solution, Colloids and Surfaces A: Physicochem. Eng. Aspects 372 (2010) 204–209. https://doi.org/10.1016/j.colsurfa.2010.10.028

[34] A.R. Rautio, P. Maki-Arvela, A. Aho, K. Eranen, K. Kordas, Chemoselective hydrogenation of citral by Pt and Pt-Sn catalysts supported on TiO_2 nanoparticles and nanowires, Catal. Today 241 (2015) 170-178. https://doi.org/10.1016/j.cattod.2013.12.052

[35] K. Iwashina, A. Iwase, Y.H. Ng, R. Amal, A. Kudo, Z-schematic water splitting into H_2 and O_2 using metal sulfide as a hydrogen-evolving photocatalyst and reduced graphene oxide as a solid-state electron mediator, J. Am. Chem. Soc. 137 (2015) 604-607. https://doi.org/10.1021/ja511615s

[36] T.-F. Yeh, J. Cihlar, C.-Y. Chang, C. Cheng, H. Teng, Roles of graphene oxide in photocatalytic water splitting, Mater. Today 16 (2013) 78-84. https://doi.org/10.1016/j.mattod.2013.03.006

[37] M.A. Nadeem, Non-noble metal photocatalysts for hydrogen production: a step ahead towards practical applications, Int. J. Petrochem. Sci. Eng. 1 (2016) 85–86. https://doi.org/10.15406/ipcse.2016.01.00015

[38] Q. Wang, T. Hisatomi, M. Katayama, T. Takata, T. Minegishi, A. Kudo, T. Yamada, K. Domen, Particulate photocatalyst sheets for Z-scheme water splitting: advantages over powder suspension and photoelectrochemical systems and future challenges, Faraday Discuss. 197 (2017) 491–504. https://doi.org/10.1039/C6FD00184J

[39] G. Darabhara, M.A. Amin, G.A.M. Mersal, E.M. Ahmed, M.R. Das, M.B. Zakaria, V. Malgras, S.M. Alsheri, Y. Yamauchi, S. Szunerits, R. Boukherroub, Reduced graphene oxide nanosheets decorated with Au, Pd and Au-Pd bimetallic nanoparticles as highly efficient catalysts for electrochemical hydrogen generation, J. Mater. Chem. A 3 (2015) 20254-20266. https://doi.org/10.1039/C5TA05730B

[40] C. Wang, D. Astruc, Nano gold plasmonic photocatalysis for organic synthesis and clean energy conversion, Chem. Soc. Rev. 43 (2014) 7188-7216. https://doi.org/10.1039/C4CS00145A

[41] S.-S. Yi, X.-B. Zhang, B.-R. Wulan, J.-M. Yan, Q. Jiang, Non-noble metals applying to solar water splitting, Energy Environ. Sci. 11 (2018) 3128-3156. https://doi.org/10.1039/C8EE02096E

[42] E. Halary-Wagner, F. Wagner, P. Hoffmann, Titanium dioxide thin-film deposition on polymer substrate by light induced chemical vapor deposition, J. Electrochem. Soc. 151 (2004) 571-576. https://doi.org/10.1149/1.1775931

[43] Y.W. Phuan, W.J. Ong, M.N. Chong, J.D. Ocon, Prospects of electrochemically synthesized hematite photoanodes for photoelectrochemical water splitting: a review, J. Photochem. Photobiol. C: Photochem. Rev. 33 (2017) 54-82. https://doi.org/10.1016/j.jphotochemrev.2017.10.001

[44] K.M.H. Young, B.M. Klahr, O. Zandi, T.W. Hamann, Photocatalytic water oxidation with hematite electrodes, Catal. Sci. Technol. 3 (2013) 1660-1671. https://doi.org/10.1039/c3cy00310h

[45] A. Wolcott, W.A. Smith, T.R. Kuykendall, Y. Zhao, J.Z. Zhang, Photoelectrochemical study of nanostructured ZnO thin films for hydrogen generation from water splitting, Adv. Funct. Mater. 19 (2009) 1849-1856. https://doi.org/10.1002/adfm.200801363

[46] E. Chorbadzhiyska, M. Mitov, G. Hristov, N. Dimcheva, L. Nalbandian, A. Evdou, Y. Hubenova, Pd-Au electrocatalysts for hydrogen evolution reaction at neutral pH, Int. J. Electrochem. 2014 (2014) 1–6. https://doi.org/10.1155/2014/239270

[47] H.W. Chen, Y. Ku, Y.L. Kuo, Effect of Pt/TiO_2 characteristics on temporal behavior of o-cresol decomposition by visible light-induced photocatalysis, Water Res. 41 (2007) 2069–2078. https://doi.org/10.1016/j.watres.2007.02.021

[48] X. Meng, L. Liu, S. Ouyang, H. Xu, D. Wang, N. Zhao, J. Ye, Nanometals for solar to chemical energy conversion: from semiconductor based photocatalysis to plasmon-mediated photocatalysis and photo-thermocatalysis, Adv. Mater. 28 (2016) 6781-6803. https://doi.org/10.1002/adma.201600305

[49] M.R. Gholipour, C.-T. Dinh, F. Beland, T.-O. Do, Nanocomposite heterojunctions as sunlight-driven photocatalysts for hydrogen production from water splitting, Nanoscale 7 (2015) 8187-8208. https://doi.org/10.1039/C4NR07224C

[50] A. Fujishima, K. Honda, S. Kikuchi, K. Kagaku Zasshi, Photosensitized electrolytic oxidation on semiconducting n-Type TiO_2 electrode, J. Chem. Soc. Japan 72 (1969) 108-113. https://doi.org/10.1246/nikkashi1898.72.108

[51] T.S. Atabaev, M.A. Hossain, D. Lee, H.K. Kim, Y.H. Hwang, Pt-coated TiO_2 nanorods for photoelectrochemical water splitting applications, Results in Physics 6 (2016) 373–376. https://doi.org/10.1016/j.rinp.2016.07.002

[52] F. Wang, Z. Zheng, F. Jia, Enhanced photoelectrochemical water splitting on Pt-loaded TiO_2 nanorods array thin film, Mater. Lett. 71 (2012) 141–144. https://doi.org/10.1016/j.matlet.2011.12.063

[53] A. Galinska, J. Walendziewski, Photocatalytic water splitting over Pt–TiO_2 in the presence of sacrificial reagents, Energy Fuels 19 (2005) 1143–1147. https://doi.org/10.1021/ef0400619

[54] J. Xiao, X. Zhang, Y. Li, A ternary g-C_3N_4/Pt/ZnO photoanode for efficient photoelectrochemical water splitting, Int. J. Hydrog. Energy 40 (2015) 9080–9087. https://doi.org/10.1016/j.ijhydene.2015.05.122

[55] N. Naseri, P. Sangpour, S.H. Mousavi, Applying alloyed metal nanoparticles to enhance solar assisted water splitting, RSC Adv. 4 (2014) 46697–46703. https://doi.org/10.1039/C4RA08216H

[56] K. Ueno, H. Misawa, Plasmon-enhanced photocurrent generation and water oxidation from visible to near-infrared wavelengths, NPG Asia Materials 5 (2013) 61-67. https://doi.org/10.1038/am.2013.42

[57] M. Haro, R. Abargues, I. Herraiz-Cardona, J. Mart'ınez-Pastor, S. Gim'enez, Plasmonic versus catalytic effect of gold nanoparticles on mesoporous TiO_2 electrodes for water splitting, Electrochim. Acta 144 (2014) 64–70. https://doi.org/10.1016/j.electacta.2014.07.146

[58] H.J. Kim, S.H. Lee, A.A. Upadhye, I. Ro, M.I. Tejedor-Tejedor, M.A. Anderson, W.B. Kim, G.W. Huber, Plasmon-enhanced photoelectrochemical water splitting with size-controllable gold nanodot arrays, ACS Nano 8 (2014) 10756–10765. https://doi.org/10.1021/nn504484u

[59] Z. Zhang, L. Zhang, M. N. Hedhili, H. Zhang, P. Wang, Plasmonic gold nanocrystals coupled with photonic crystal seamlessly on TiO_2 nanotube photoelectrodes for efficient visible light photoelectrochemical water splitting, Nano Lett. 13 (2013) 14–20. https://doi.org/10.1021/nl3029202

[60] X. Zhang, Y. Liu, S.-T. Lee, S. Yang, Z. Kang, Coupling surface plasmon resonance of gold nanoparticles with slow-photon-effect of TiO_2 photonic crystals for synergistically enhanced photoelectrochemical water splitting, Energy Environ. Sci. 7 (2014) 1409–1419. https://doi.org/10.1039/c3ee43278e

[61] F. Su, T. Wang, R. Lv, J. Zhang, P. Zhang, J. Lu, J. Gong, Dendritic Au/TiO_2 nanorod arrays for visible-light driven photoelectrochemical water splitting, Nanoscale 5 (2013) 9001-9010. https://doi.org/10.1039/c3nr02766j

[62] J.Y. Choi, Y. Hoon Sung, H.J. Choi, Y. Doo Kim, D. Huh, H. Lee, Fabrication of Au nanoparticle-decorated TiO_2 nanotube arrays for stable photoelectrochemical water splitting by two-step anodization, Ceram. Int. 43 (2017) 14063–14067. https://doi.org/10.1016/j.ceramint.2017.07.141

[63] P. Peerakiatkhajohn, T. Butburee, J.H. Yun, H. Chen, R.M. Richards, L. Wang, A hybrid photoelectrode with plasmonic $Au@TiO_2$ nanoparticles for enhanced photoelectrochemical water splitting, J. Mater. Chem. A 3 (2015) 20127–20133. https://doi.org/10.1039/C5TA04137F

[64] J. Chen, M. Yu, Y. Wang, S. Shen, M. Wang, L. Guo, $Au@SiO_2$ core/shell nanoparticle-decorated TiO_2 nanorod arrays for enhanced photoelectrochemical water splitting, Chin. Sci. Bull. 59 (2014) 2191–2198. https://doi.org/10.1007/s11434-014-0188-7

[65] L. Wang, X. Zhou, N.T. Nguyen, P. Schmuki, Plasmon-enhanced photoelectrochemical water splitting using Au nanoparticles decorated on hematite nanoflake arrays, ChemSusChem 8 (2015) 618–622. https://doi.org/10.1002/cssc.201403013

[66] M. Gholami, M. Qorbani, O. Moradlou, N. Naseri, A.Z. Moshfegh, Optimal Ag_2S nanoparticle incorporated TiO_2 nanotube array for visible water splitting, RSC Adv. 4 (2014) 7838- 7844. https://doi.org/10.1039/c3ra44898c

[67] Y. Wei, L. Ke, J. Kong, H. Liu, Z. Jiao, X. Lu, H. Du, X.W. Sun, Enhanced photoelectrochemical water-splitting effect with a bent ZnO nanorod photoanode decorated with Ag nanoparticles, Nanotechnology 23 (2012) 235401-235409. https://doi.org/10.1088/0957-4484/23/23/235401

[68] J. Wu, S. Lu, D. Ge, L. Zhang, W. Chen, H. Gu, Photocatalytic properties of Pd/TiO_2 nanosheets for hydrogen evolution from water splitting, RSC Adv. 6 (2016) 67502-67508. https://doi.org/10.1039/C6RA10408H

[69] N.K. Allam, A.J. Poncheri, M.A. El-Sayed, Vertically oriented Ti–Pd mixed oxynitride nanotube arrays for enhanced photoelectrochemical water splitting, ACS Nano 5 (2011) 5056–5066. https://doi.org/10.1021/nn201136t

[70] G. Hitoki, T. Takata, J.N. Kondo, M. Hara, H. Kobayashi, K. Domen, (Oxy) nitrides as new photocatalysts for water splitting under visible light irradiation, J. Electrochem. Soc. 70 (2002) 463–465. https://doi.org/10.5796/electrochemistry.70.463

[71] C. Hou, M. Zhang, T. Kasama, C. Engelbrekt, L. Zhang, H. Wang, Q. Chi, Reagent-free synthesis and plasmonic antioxidation of unique nanostructured metal-metal oxide core shell microfibers, Adv. Mater. 28 (2016) 4097-4104. https://doi.org/10.1002/adma.201505990

[72] S. Zhang, B. Peng, S. Yang, H.G. Wang, H. Yu, Y. Fang, F. Peng, Non-noble metal copper nanoparticles decorated TiO_2 nanotube arrays with plasmon-enhanced photocatalytic hydrogen evolution under visible light, Int. J. Hydrog. Energy 40 (2015) 303-310. https://doi.org/10.1016/j.ijhydene.2014.10.122

[73] G. Xiong, R. Shao, T.C. Droubay, A.G. Joly, K.M. Beck, S.A. Chambers, Photoemission electron microscopy of TiO_2 anatase films embedded with rutile nanocrystals, Adv. Funct. Mater. 17 (2007) 2133-2138. https://doi.org/10.1002/adfm.200700146

[74] F. Wang, R.J. Wong, J.H. Ho, Y. Jiang, R. Amal, Sensitization of Pt/TiO_2 using plasmonic Au nanoparticles for hydrogen evolution under visible-light irradiation, ACS Appl. Mater. Interfaces 9 (2017) 30575–30582. https://doi.org/10.1021/acsami.7b06265

[75] A.A. Melvin, K. Illath, T. Das, T. Raja, S. Bhattacharyya, C.S. Gopinath, M-Au/TiO_2 (M=Ag, Pd and Pt) Nanophotocatalyst for overall solar water splitting: role of interfaces, Nanoscale 7 (2015) 13477-13488. https://doi.org/10.1039/C5NR03735B

[76] Y. Lu, J. Zhang, L. Ge, C. Han, P. Qiu, S. Fang, Synthesis of novel Au-Pd nanoparticles decorated one-dimensional ZnO nanorod arrays with enhanced photoelectrochemical water splitting activity, J. Colloid Interface Sci. 483 (2016) 146–153. https://doi.org/10.1016/j.jcis.2016.08.022

[77] R.S. Moakhar, A. Kushwaha, M. Jalali, G.K.L. Goh, A. Dolati, M. Ghorbani, Enhancement in solar driven water splitting by Au-Pd nanoparticle decoration of electrochemically grown ZnO nanorods, J. Appl. Electrochem. 46 (2016) 819-827. https://doi.org/10.1007/s10800-016-0981-x

[78] Y. Li, Z. Liu, Y. Wang, Z. Liu, J. Han, J. Ya, $ZnO/CuInS_2$ core/shell heterojunction nanoarray for photoelectrochemical water splitting, Int. J. Hydrog. Energy 37 (2012) 15029-15037. https://doi.org/10.1016/j.ijhydene.2012.07.117

[79] J. Han, Z. Liu, K. Guo, J. Ya, Y. Zhao, X. Zhang, T. Hong, J. Liu, High-efficiency $AgInS_2$-modified ZnO nanotube array photoelectrodes for all-solid-state hybrid solar cells, ACS Appl. Mater. Interfaces 6 (2014) 17119-17125. https://doi.org/10.1021/am5047813

[80] M. Law, L.E. Greene, J.C. Johnson, R. Saykally, P. Yang, Nanowire dye-sensitized solar cells, Nat. Mater. 4 (2005) 455-459. https://doi.org/10.1038/nmat1387

[81] Y. Zhang, Y. Zhang, Y. Guo, L. Wu, Y. Liu, L. Song, Synthesis of Au-Pd nanoparticle-decorated graphene-coated ZnO nanorod arrays with enhanced photoelectrochemical performance and stability, RSC Adv. 9 (2018) 2666–2672. https://doi.org/10.1039/C8RA09028A

[82] K.K. Patra, C.S. Gopinath, Bimetallic and plasmonic Ag-Au on TiO_2 for solar water splitting: an active nanocomposite for entire visible-light-region absorption, ChemCatChem 8 (2016) 3294–3311. https://doi.org/10.1002/cctc.201600937

[83] J. Liu, F. Chen, Plasmon enhanced photoelectrochemical activity of Ag-Cu nanoparticles on TiO_2/Ti substrates, Int. J. Electrochem. Sci. 7 (2012) 9560-9572.

[84] S. Kim, J.-M. Kim, J.-E. Park, J.-M. Nam, Non noble metal based plasmonic nanomaterials: recent advances and future perspectives, Adv. Mater. 30 (2018) 1704528-1704551. https://doi.org/10.1002/adma.201704528

[85] N.M. Mohamed, R. Bashiri, F.K. Chong, S. Sufian, S. Kakooei, Photoelectrochemical behavior of bimetallic Cu–Ni and monometallic Cu, Ni doped TiO_2 for hydrogen production. Int. J. Hydrog. Energy 40 (2015) 14031–14038. https://doi.org/10.1016/j.ijhydene.2015.07.064

[86] H. Tian, S.Z. Kang, X. Li, L. Qin, M. Ji, J. Mu, Fabrication of an efficient noble metal-free TiO_2 based photocatalytic system using Cu–Ni bimetallic deposit as an active center of H_2 evolution from water, Sol. Energy Mater. Sol. Cells 134 (2015) 309–317. https://doi.org/10.1016/j.solmat.2014.12.016

[87] M.M. Momeni, Y. Ghayeb, M. Mahvari, Study of photoelectrochemical water splitting using films based on deposited TiO_2 nanotubes, Appl. Phys. A 124 (2018) 586-597. https://doi.org/10.1007/s00339-018-2009-3

[88] B.Y. Cheng, J.S. Yang, H.W. Cho, J.J. Wu, Fabrication of efficient $BiVO_4$-TiO_2 heterojunction photoanode for photoelectrochemical water oxidation, ACS Appl. Mater. Interfaces 8 (2016) 20032-20039. https://doi.org/10.1021/acsami.6b05489

[89] F. Cao, W. Tian, L. Li, Ternary non-noble metal zinc-nickel-cobalt carbonate hydroxide co-catalysts toward highly efficient photoelectrochemical water splitting, J. Mater. Sci. Technol. 34 (2017) 899–904. https://doi.org/10.1016/j.jmst.2017.11.054

[90] X. Han, Y. Wei, J. Su, Y. Zhao, Low-cost oriented hierarchical growth of BiVO4/rGO/Ni-Fe nanoarrays photoanode for photoelectrochemical water splitting, ACS Sustainable Chem. Eng. 6 (2018) 14695−14703. https://doi.org/10.1021/acssuschemeng.8b03259

[91] S. Oh, S. Jung, Y.H. Lee, J.T. Song, T.H. Kim, D.K. Nandi, S.-H. Kim, J. Oh, Hole-selective CoO_x/SiO_x/Si heterojunctions for photoelectrochemical water splitting, ACS Catal. 8 (2018) 9755-9764. https://doi.org/10.1021/acscatal.8b03520

[92] S.R. Vaddipalli, S.R. Sanivarapu, S. Vengatesan, J.B. Lawrence, M. Eashwar, G. Sreedhar, Heterostructured Au NPs/CdS/LaBTC MOFs photoanode for efficient photoelectrochemical water splitting: stability enhancement via CdSe QDs to 2D-CdS nanosheets transformation, ACS Appl. Mater. Interfaces 8 (2016) 23049−23059. https://doi.org/10.1021/acsami.6b06851

Keyword Index

About the Editors

Dr. Inamuddin is currently working as Assistant Professor in the Chemistry Department, Faculty of Science, King Abdulaziz University, Jeddah, Saudi Arabia. He is a permanent faculty member (Assistant Professor) at the Department of Applied Chemistry, Aligarh Muslim University, Aligarh, India. He obtained Master of Science degree in Organic Chemistry from Chaudhary Charan Singh (CCS) University, Meerut, India, in 2002. He received his Master of Philosophy and Doctor of Philosophy degrees in Applied Chemistry from Aligarh Muslim University (AMU), India, in 2004 and 2007, respectively. He has extensive research experience in multidisciplinary fields of Analytical Chemistry, Materials Chemistry, and Electrochemistry and, more specifically, Renewable Energy and Environment. He has worked on different research projects as project fellow and senior research fellow funded by University Grants Commission (UGC), Government of India, and Council of Scientific and Industrial Research (CSIR), Government of India. He has received Fast Track Young Scientist Award from the Department of Science and Technology, India, to work in the area of bending actuators and artificial muscles. He has completed four major research projects sanctioned by University Grant Commission, Department of Science and Technology, Council of Scientific and Industrial Research, and Council of Science and Technology, India. He has published 147 research articles in international journals of repute and eighteen book chapters in knowledge-based book editions published by renowned international publishers. He has published 60 edited books with Springer (U.K.), Elsevier, Nova Science Publishers, Inc. (U.S.A.), CRC Press Taylor & Francis Asia Pacific, Trans Tech Publications Ltd. (Switzerland), IntechOpen Limited (U.K.), and Materials Research Forum LLC (U.S.A). He is a member of various journals' editorial boards. He is also serving as Associate Editor for journals (Environmental Chemistry Letter, Applied Water Science and Euro-Mediterranean Journal for Environmental Integration, Springer-Nature), Frontiers Section Editor (Current Analytical Chemistry, Bentham Science Publishers), Editorial Board Member (Scientific Reports-Nature), Editor (Eurasian Journal of Analytical Chemistry), and Review Editor (Frontiers in Chemistry, Frontiers, U.K.) He is also guest-editing various special thematic special issues to the journals of Elsevier, Bentham Science Publishers, and John Wiley & Sons, Inc. He has attended as well as chaired sessions in various international and national conferences. He has worked as a Postdoctoral Fellow, leading a research team at the Creative Research Initiative Center for Bio-Artificial Muscle, Hanyang University, South Korea, in the field of renewable energy, especially biofuel cells. He has also worked as a Postdoctoral Fellow at the Center of Research Excellence in Renewable Energy, King Fahd University of Petroleum and Minerals, Saudi Arabia, in the field of polymer electrolyte membrane fuel cells and computational fluid dynamics of polymer electrolyte membrane fuel cells. He is a life member of the Journal of the Indian Chemical Society. His research interest includes ion exchange materials, a sensor for heavy metal ions, biofuel cells, supercapacitors and bending actuators.

Dr. Rajender Boddula is currently working with Chinese Academy of Sciences-President's International Fellowship Initiative (CAS-PIFI) at National Center for Nanoscience and Technology (NCNST, Beijing). He obtained Master of Science in Organic Chemistry from Kakatiya University, Warangal, India, in 2008. He received his Doctor of Philosophy in Chemistry with the highest honours in 2014 for the work entitled "Synthesis and Characterization of Polyanilines for Supercapacitor and Catalytic Applications" at the CSIR-Indian Institute of Chemical Technology (CSIR-IICT) and Kakatiya University (India). Before joining National Center for Nanoscience and Technology (NCNST) as CAS-PIFI research fellow, China, worked as senior research associate and Postdoc at the Aligarh Muslim University (AMU, India) and National Tsing-Hua University (NTHU, Taiwan) respectively in the fields of bio-fuel and CO_2 reduction applications. His academic honors include University Grants Commission National Fellowship and many merit scholarships, study-abroad fellowships from Australian Endeavour Research Fellowship, and CAS-PIFI. He has published many scientific articles in international peer-reviewed journals and has authored around twenty book chapters, and he is also serving as an editorial board member and a referee for reputed international peer-reviewed journals. He has published edited books with Springer (UK), Elsevier, Materials Science Forum LLC (USA) and CRC Press Taylor & Francis group. His specialized areas of research are energy conversion and storage, which include sustainable nanomaterials, graphene, polymer composites, heterogeneous catalysis for organic transformations, environmental remediation technologies, photoelectrochemical water-splitting devices, biofuel cells, batteries and supercapacitors.

Dr. Mohammad Faraz Ahmer is presently working as Assistant Professor in the Department of Electrical Engineering, Mewat Engineering College, Nuh Haryana, India, since 2012 after working as Guest Faculty in University Polytechnic, Aligarh Muslim University Aligarh, India, during 2009-2011. He completed M.Tech. (2009) and Bachelor of Engineering (2007) degrees in Electrical Engineering from Aligarh Muslim University, Aligarh in the first division. He obtained a Ph.D. degree in 2016 on his thesis entitled "Studies on Electrochemical Capacitor Electrodes". He has published six research papers in reputed scientific journals. He has edited two books with Materials Research Forum, U.S.A. His scientific interests include electrospun nano-composites and supercapacitors. He has presented his work at several conferences. He is actively engaged in searching of new methodologies involving the development of organic composite materials for energy storage systems.

Prof. Abdullah M. Asiri is the Head of the Chemistry Department at King Abdulaziz University since October 2009 and he is the founder and the Director of the Center of Excellence for Advanced Materials Research (CEAMR) since 2010 till date. He is the Professor of Organic Photochemistry. He graduated from King Abdulaziz University (KAU) with B.Sc. in Chemistry in 1990 and a Ph.D. from University of Wales, College of Cardiff, U.K. in 1995. His research interest covers color chemistry, synthesis of novel photochromic and thermochromic systems, synthesis of novel coloring matters and dyeing of textiles, materials chemistry, nanochemistry and nanotechnology, polymers and plastics. Prof. Asiri is the principal supervisors of more than 20 M.Sc. and six Ph.D. theses. He is the main author of ten books of different chemistry disciplines. Prof. Asiri is the Editor-in-Chief of King Abdulaziz University Journal of Science. A major achievement of Prof. Asiri is the discovery of tribochromic compounds, a new class of compounds which change from slightly or colorless to deep colored when subjected to small pressure or when grind. This discovery was introduced to the scientific community as a new terminology published by IUPAC in 2000. This discovery was awarded a patent from European Patent office and from UK patent. Prof. Asiri involved in many committees at the KAU level and on the national level. He took a major role in the advanced materials committee working for KACST to identify the national plan for science and technology in 2007. Prof. Asiri played a major role in advancing the chemistry education and research in KAU. He has been awarded the best researchers from KAU for the past five years. He also awarded the Young Scientist Award from the Saudi Chemical Society in 2009 and also the first prize for the distinction in science from the Saudi Chemical Society in 2012. He also received a recognition certificate from the American Chemical Society (Gulf region Chapter) for the advancement of chemical science in the Kingdome. He received a Scopus certificate for the most publishing scientist in Saudi Arabia in chemistry in 2008. He is also a member of the editorial board of various journals of international repute. He is the Vice- President of Saudi Chemical Society (Western Province Branch). He holds four USA patents, more than one thousand publications in international journals, several book chapters and edited books.

www.ingramcontent.com/pod-product-compliance
Lightning Source LLC
Chambersburg PA
CBHW071205210326
41597CB00016B/1684